# MAP PROJECTIONS

# MAP PROJECTIONS

## FOR GEODESISTS, CARTOGRAPHERS AND GEOGRAPHERS

BY

PETER RICHARDUS, Ph. D.

AND

RON K. ADLER, D. Sc.

1972

NORTH-HOLLAND PUBLISHING COMPANY – AMSTERDAM · LONDON
AMERICAN ELSEVIER PUBLISHING CO., INC. – NEW YORK

*Library of Congress Catalog Card Number: 79-182493*
*North-Holland ISBN: 0 7204 5007 1*
*American Elsevier ISBN: 0 444 10362 7*

91 graphs and illustrations, 8 tables

PUBLISHERS:
NORTH-HOLLAND PUBLISHING COMPANY – AMSTERDAM
NORTH-HOLLAND PUBLISHING COMPANY, LTD. – LONDON

SOLE DISTRIBUTORS FOR THE U.S.A. AND CANADA:
AMERICAN ELSEVIER PUBLISHING COMPANY, INC.
52 VANDERBILT AVENUE, NEW YORK, N.Y. 10017

PRINTED IN BELGIUM

# PREFACE

The idea to write this introduction to map projections was born in 1969 in the Department of Geodetic Science of the Ohio State University, Columbus Ohio, U.S.A. when one of the authors had to take over lectures on this subject from the other.

There existed agreement in the opinions of the authors – and they know themselves to be supported by many colleagues – that most books on map projections are either too simple of conception, or too involved, carrying the subject matter far beyond the knowledge required of geodesists studying at a Master's degree level. Also, relevant publications of various nature are numerous and diverse.

This book is to be an intermediate. After a description of general features presupposing a knowledge of elementary spherical trigonometry, it employs mathematics up to a level of ordinary differential and integral calculus. It leads the way directly to the design of general computer programmes for the calculation and/or plotting of geographic grids in the common projections of both the terrestrial and lunar ellipsoids. At the same time an endeavour is made to provide a solid basis for further study by a more interested reader.

Until a decade ago cartographers and geographers generally did not receive sufficient mathematical instruction for a proper understanding of map projections as they are treated in this book. However, for many disciplines this instruction at the present seems to have been greatly improved under the impact of modern developments in geosciences. A suggested course of reading for the specific type of student is given with the list of contents.

A remark should be made about the notation. In the literature of science generally no notation can be found less consistent and diverse than that used in the literature of map projections. In this text the notation of parameters as it is used in common geodetic practice has been followed closely, with corresponding capitals in the projection surface. It is not ideal, but it worked satisfactorily. The few proverbial exceptions have been defined locally in the text.

It is hoped that the combined experience of the authors in geodesy and cartography, in the academic and professional practice, has produced a text useful to students and colleagues interested in the application of projections in terrestrial and lunar surveying and mapping.

*October 1971*

PETER RICHARDUS, RON K. ADLER

# CONTENTS

Suggested simplified course of reading:
Chapters 1 and 2
Chapter 3 sections 1; 2; 3.1; 3.2 (up to form. 3.32 incl.) 3.4 and 3.5
Chapter 4
Chapter 5 sections 1.1; 1.2; 3.1; 3.2 (1 standard parallel) 3.3; 3.4; 3.5
Chapter 6 sections 1.1; 3.1; 3.2 (1 standard parallel); 3.3
Chapter 7

## CHAPTER 1

INTRODUCTION

## 1.1 The basic problem of map projections

The basic problem of map projections is the representation of a curved surface in a plane. Applicationally, it is often the problem of representing the earth or the moon on a flat map. The figure of the earth is usually represented by a solid of revolution, either the ellipsoid or the sphere, as the case may be, which is regarded as a reference surface to which all physical points are related. These points may be situated on dry land surface of the earth, on the surface of seas, oceans and lakes or below such water surfaces on the earth. It is simple to imply that the reference surface is the representation of the *mean* sea level and its continuation under dry land or over dry land depressions. This implication is not, strictly speaking, true since the figure of the earth is truly represented only by an equipotential surface at the mean sea level, called the geoid, and such an equipotential surface is irregular, or undulating, impossible to express by a rigorous mathematical formula. The determination of the geoid and the choice of a regularly shaped reference surface which would best represent the figure of the earth is one of the tasks of geodesy. In recent years this task has been extended to the moon, and thus, we have a new branch of science dealing with the size and shape of the moon called selenodesy.

It will, however, be assumed that the currently determined reference surfaces for the earth and moon are an ellipsoid and a sphere of known parameters respectively, and thus, the problem of map projections is confined to the representation of the ellipsoidal and spherical surfaces on a plane.

It may be further stated that there is no perfect solution to the problem, and this may be readily surmised by trying to apply an orange peel to a flat table surface. To achieve a continuous contact between the two surfaces, the orange peel would have to be distorted by stretching, shrinking or tearing and this, while being a gross oversimplification of the problem of map projections, is nevertheless a good illustration of the impossibility of a perfect solution (Section 2.3.2). One may argue that the whole problem is totally irrelevant, since it is possible to limit ourselves to three dimensional models of the ellipsoid or the sphere, either physically constructed, as in the case of the globe, or mathematically expressed, as done by geodesists. Theoretically, such an argument is perfectly valid and the desire for a representation on a plane surface is purely one of convenience. There are many valid reasons for desiring such a convenience, the obvious ones being that a flat map is easier to produce and handle than a globe or a scaled portion of a curved surface and the computations on a plane are much easier than computations on an ellipsoid or a sphere, even in the age of an electronic computer.

We may, thus, conclude that all representations of curved surfaces on a plane involve "stretching" or "shrinking" resulting in distortions or "tearing" resulting in interruptions. Different techniques of representation are applied to achieve representations which possess certain properties favorable for the specific purpose at hand, and considering the size of the area to be represented. The technique of representation is commonly called map projection, although this term is not to be taken literally, since not all representations are produced for mapping and certainly not all representations are achieved by means of geometric projection.

The term map projection or projection will be used throughout this book, but one should remember that the final product in each transformation from the original curved surface to the plane surface is a representation achieved through a specific technique.

## 1.2   Purposes and methods of projections

The representation of the earth or lunar surface on a plane may be required for the purpose of expressing the position of discrete points on the original surface in a plane coordinate system and the computation of distances and directions within a system of such discrete points. This is usually of primary interest to the surveyor and mapper, who deals with areas of usually limited extent within his scope of activity.

Representations of curved surfaces on a plane are also made for the purpose of graphical presentations which are of primary interest to the geographer as aids to the studies of topography, habitation, climatology, vegetation etc., dealing usually with areas of greater extent.

Three principal cartographic criteria are applied to the evaluation of map projection properties:

(a)   Equidistance – correct representation of distances.

(b)   Conformality (orthomorphism) – correct representation of shapes.

(c)   Equivalency – correct representation of areas.

The above three criteria are basic and mutually exclusive, other properties being of secondary nature. Considering this, it should be stressed that there is no ideal representation, only the best representation for a given purpose.

The methods of projection or transformation may be classified as

(a)   Direct projection from the ellipsoidal to the projection surface.

(b)   Double projection involving a transformation from ellipsoidal to spherical surface and then from the spherical to the projection surface.

We have thus, two kinds of datum surfaces – *ellipsoid* and *sphere*. There are three kinds of projection surfaces – *plane, cone* and *cylinder*, the latter two being developable into a plane.

The transformation from datum to projection surface may be *geometrical, semi-geometrical* or *mathematical* in nature. Very few of the transformations are truly perspective projections in a geometrical sense.

It is convenient to define a map projection as a systematic arrangement of intersecting lines on a plane, that represent and have a one-to-one correspondence to the meridians and parallels on the datum surface. The arrangement follows some consistent principle in order to fulfill certain required conditions. Each set of new conditions results in a different map projection and thus, potentially there exists an unlimited number of map projections. In practice, however, the three above mentioned principal cartographic criteria are applied with a rather limited number of other conditions resulting in a number of some two hundred odd map projections created for specific purposes (see Maling's list of projections [14] many of which are of no general interest).

## 1.3   Classification of map projections

The classification of map projections should follow a standard pattern so that any regular projection (non-conventional) can be described by a set of criteria and conversely a set of criteria will define a regular projection. Thus, a classification scheme may follow a number of criteria subdivided into *classes* and *varieties* as suggested by Goussinsky [8].

*Classes* may be considered as different points of view. These points of view are not mutually exclusive. An example of such non-exclusiveness is the consideration of a person as, say, a human being, Frenchman, red haired, Protestant, vegetarian etc. This corresponds respectively to classes of species, nationality, hair color, religion, type of food consumed etc., none of which are mutually exclusive.

*Varieties* are the accepted or existing subdivisions within each class and they are mutually exclusive since a person would have a certain one nationality, one hair color, one religion etc. To facilitate the setting up of a classification scheme for map projections composed of classes and varieties certain specific factors should be considered.

(a)   The projected object or datum surface.

(b)   The projection surface upon which the datum surface is projected.

(c)   The projection or representation per se.

The projection surface is considered as the extrinsic problem and the process of projection or representation as the intrinsic problem.

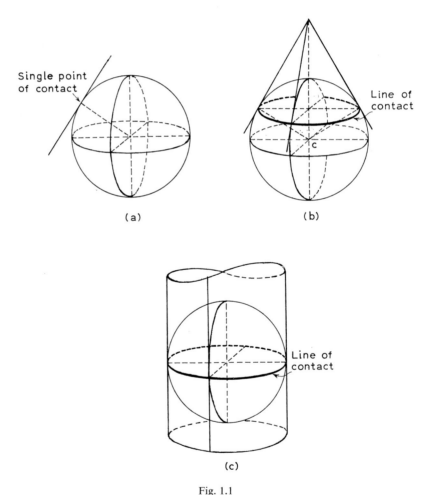

Fig. 1.1
The three basic projection surfaces. (a) Plane, (b) cone, (c) cylinder, showing the respective extent of contact with the datum surface

### 1.3.1 *Extrinsic problem*

This problem involves the consideration of the properties of the projection surface relative to the datum surface giving rise to three classes :

I. *Nature* of the projection surface defined as geometric figure.

II. *Coincidence* or contact of the projection surface with the datum surface.

III. *Position* or alignement of the projection surface with relation to the datum surface.

*Class I* may be further subdivided into three varieties, each representing one of the basic projection surfaces, namely, the *plane*, the *cone* and the *cylinder*. The simplest of these projection surfaces is the plane, which when tangent to the datum surface would have a single point of contact, this being also the center of the area of minimum distortion. The cone and the cylinder, which are both developable into a plane were introduced with view to increase the extent of contact and, consequently the area of minimum distortion (see Fig. 1.1).

*Class II* is thus further subdivided into three varieties representing the three types of coincidence between the datum and projection surface, namely, *tangent*, *secant* and *polysuperficial*. It may easily be surmised that tangency between the datum and projection surfaces results in a point or line contact, the former in the case of projection surface being a plane and the latter in the case of projection surface being either a cone or a cylinder. To increase the contact between the surfaces and thus, also the area of minimum distortion, the secant case has been introduced resulting in a line of contact instead of a point when the projection surface is a plane, and in two lines of contact when the projection is either a cone or a cylinder. A still further increase of contact and consequently in areas of minimum distortion is achieved by employment of polysuperficiality, or in other words, a series of successive projection surfaces. A series of successive planes would produce a polyhedric (multiple plane) projection, a series of cones produces a polyconic and a series of cylinders a polycylindrical projection. This is illustrated in Figure 1.2. The development of the cones in the polyconic projection is shown in Figure 1.2d.

*Class III* is subdivided into three varieties representing the three basic positions or alignments of the projection surface relative to the datum surface, namely, *normal*, *transverse* and *oblique*. If the purpose of the projection is to represent a limited area of the datum surface, it is advantageous to achieve the minimum of distortion in that particular area. This is possible through varying the attitude of the projection surface. If the axis of symmetry of the projection surface coincides with the rotational axis of the ellipsoid or the sphere, the

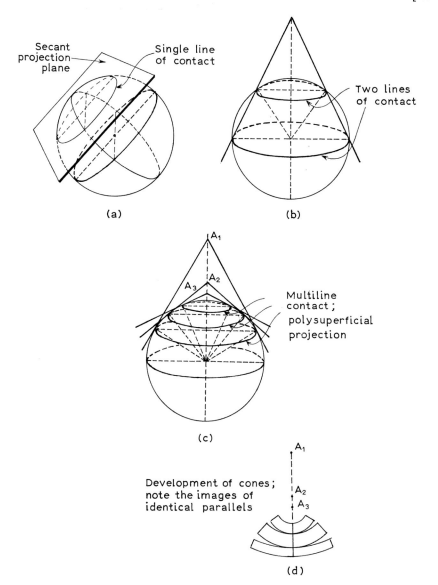

Fig. 1.2

Increase of contact between the projection and datum surfaces. (a) A single line contact when projection plane is secant to datum surface, (b) a two line contact when the projection surface is secant and, (c) multiline contact when the projection is polysuperficial. Compare Figure 1.1

normal case is obtained. With the axis of symmetry perpendicular to the axis of rotation, we have the transverse case and the other attitudes of the axis of symmetry result in oblique cases (see Fig. 1.3).

A section through the apex $A$ of the cone and the centre $C$ in Figure 1.1b is shown in Figure 1.4a. The angle $B'CD\,(\delta)$ determines the position of the circle through $B$ and $B'$, the circumference of this circle being $2\pi R \cos \delta$, where $CB' = R$. When the conical surface is developed into the plane by dissecting it along $AB$ then the image of the tangential circle (with length $2\pi R \cos \delta$)

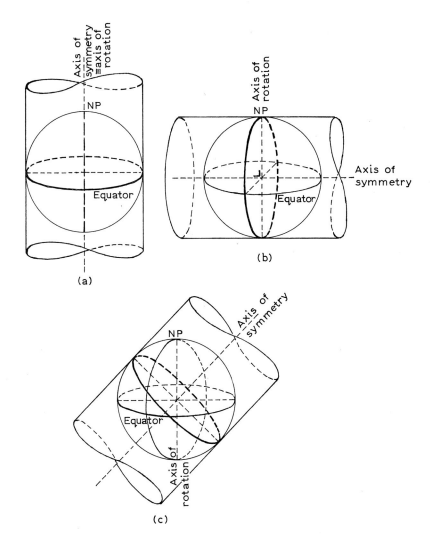

Fig. 1.3
The attitude of the projection surface (in this case a cylinder). (a) *Normal* – contact along the equator, (b) *transverse* – contact along a selected meridian, (c) *oblique* – contact along a selected great circle

becomes a part of the circumference $2\pi R \cot \delta$ of a circle with a radius equal to $AB = R \cot \delta$.

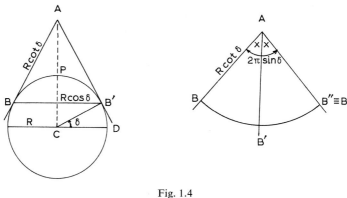

Fig. 1.4

(a)                   (b)

Development of a conical projection surface

The proportion $2\pi R \cos \delta / 2\pi R \cot \delta = \sin \delta$ is called the constant of the cone. The extreme values of this constant are $\sin \delta = 0$ or $\delta = 0$, indicating that the apex $A$ is at infinity, transforming the cone into a cylinder; and $\sin \delta = 1$ or $\delta = 90°$, indicating that the point $P$ and the apex are in coincidence, transforming the cone into a plane tangent at $P$.

In the normal position of the cone the angle $\delta$ is equal to the latitude $\varphi$ of the parallel circle $BB'$. The constant then is $\sin \varphi_{BB'}$.

As a summary of the three classes considered within the extrinsic problem it may be stated that the efforts of cartographers are directed towards increasing the area of minimum distortion by selecting the most favorable variety from each class for the specific purpose of the projection.

### 1.3.2 *Intrinsic problem*

This problem involves the consideration of a projection from the point of view of its cartographic properties and the mode of generation, giving rise to the following two classes:

  IV.   Properties

  V.   Generation

*Class IV* is further subdivided into 3 mutually exclusive varieties representing the three basic cartographic criteria according to which properties are evaluated: *equidistance, conformality* (orthomorphism) and *equivalency* (equality of areas).

Equidistance means that there is a correct representation of a distance between two points on the datum surface and the corresponding points on the projection surface, so that the scale is maintained along the lines connecting a pair of points. This is, however, limited to certain specified points and is by no means a general property between points on the two surfaces.

Equivalence of areas means that areas of figures represented are retained, but at the expense of shapes and angles which are in such case deformed.

Conformality means retention of shape or form, and thus also retention of angles (directions), this property being limited to differentially close points and certainly not valid for areas of significant dimensions.

*Class V* is further subdivided into 3 mutually exclusive varieties representing the three principal modes of generation of projections or representations. The necessity or the desire to obtain certain properties in projections or representations led to the evolution of the various modes of generation either through a purely geometric or perspective projection technique or through a process only partly projective where only one family of lines is projected by a pencil of planes all passing through the axis of symmetry of the projection surface. There are some representations, however, which are completely free of the projection operation, i.e. no rays are involved and the representation is achieved by a convention followed by a purely mathematical process. There are thus in Class V three varieties, namely, *geometric*, *semi-geometric* and *conventional*.

It is important to stress the fact that the number of conventional representations is infinite. These representations are subject to empirical or posterior classifications only since in this case we are unable to answer questions pertaining to the geometric form of the projection surface or its position relative to the datum surface.

### 1.3.3  *The classification scheme*

The previously discussed classes and varieties may be arranged as follows:

| | Classes | Varieties | | |
|---|---|---|---|---|
| Projection surface (extrinsic problem) | I. Nature | Plane | Conical | Cylindrical |
| | II. Coincidence | Tangent | Secant | Polysuperficial |
| | III. Position | Normal | Transverse | Oblique |
| The projection itself (intrinsic problem) | IV. Properties | Equidistant | Equivalent | Conformal |
| | V. Generation | Geometric | Semi-geometric | Conventional |

A regular (non-conventional) projection may be described by a set of varieties, one from each class; and conversely, a set of varieties, one from each class, defines a regular projection.


## 1.4   Datum surfaces and coordinate systems

The datum surface of the earth is usually an ellipsoid of revolution, although sometimes it is approximated by a sphere.

There are two ways of representing the datum surface on a plane or a surface developable into a plane:

(a)   The direct method, whereby a transformation is achieved directly from the ellipsoid to the projection surface.

(b)   The indirect method, whereby a transformation is first achieved from the ellipsoid to the sphere as an intermediate surface and then from the sphere to the projection surface. This method is known as a double projection.

The ellipsoid can be projected onto a sphere with relatively little distortion. Furthermore, the representation of the ellipsoid on a sphere may be sufficient in itself if the purpose of the projection is to perform computations rather than to achieve a graphical representation. This aspect used to be very important from the geodesist's point of view, since computations of geodetic networks could be performed on the sphere using simple formulas of spherical trigonometry before the advent of the electronic computer. Today, when computers are almost universally available, this advantage to the geodesist is very much diminished. However, for the purpose of graphical representation or for the purpose of operating with plane coordinates in surveying, the transformation from the sphere to the plane, is relatively a much simpler proposition than the direct transformation from the ellipsoid to the plane.

It could be safely stated that while the sphere to plane transformation is very useful for learning and understanding the various types of non-conventional projections, its practical value is today almost insignificant since the computer eliminates the computational difficulties and an automatic plotter is usually capable of efficiently presenting the computer output in graphical form when necessary.

Coordinate systems are necessary for the expression of position of points upon the various surfaces, be it an ellipsoid, a sphere or a plane. For the ellipsoid or the sphere the system of longitude and latitude expressed in degrees, minutes and seconds of arc is universally accepted (see Chapter 2). For the plane a system of rectangular $X$ and $Y$ coordinates, sometimes referred to as Northings and Eastings is usually applicable.

The graphical presentation of coordinate systems assumes the form of grids formed by regularly spaced lines of longitude (meridians) and latitude (parallels) or Eastings and Northings.

It is convenient to consider the problems of map projections as the transformation of grids from the ellipsoidal or spherical surface onto a gridded plane surface.

## 1.5   Summary of representation problems

### 1.5.1   *Non-applicability of an ellipsoid (or sphere) to a plane*

Applicable surfaces are those where one can be produced by bending the other without stretching or tearing. It is impossible to devise a projection which would be an *undeformed* representation of the earth or the moon. The mathematical proof of the inapplicability of spherical and planed surfaces is given by Craig [5].

The cartographer strives to evaluate the existing projections in order to choose the one most suitable for the given purpose, on the basis of the projection's characteristics.

### 1.5.2   *Definition of representation of one surface upon another*

Given two arbitrary surfaces defined by their respective parametric equations, every one-to-one correspondence between points on the datum surface and points on the projection surface is the representation of the datum surface upon the projection surface. The one-to-one correspondence exists if the following two conditions are satisfied:

(a)   Every point of the datum surface has a corresponding point on the projection surface, thus the one-to-one correspondence is the correspondence of points and not of lines or figures.

(b)   The correspondence must be unique and reciprocal, meaning that every point on the datum surface has only one corresponding point on the projection surface and vice versa.

### 1.5.3   *The investigation of projection properties*

The properties of the projection are investigated by comparison with the properties of the datum, bearing in mind that a perfect representation is impossible. Thus, several factors should be considered:

(a) Certain properties of the datum may remain unchanged in the projection.

(b) Certain other properties of the datum may change in the projection or disappear completely.

(c) The projection may acquire properties which did not exist in the datum.

The theory of map projections is concerned with the examination of the unchanged properties of the datum and the investigation of the changed properties of the datum in the map projection. The qualitative and quantitative investigation of such changes is known as the *theory of deformations* (distortions).

Bearing in mind that irrespective of how the spherical datum surface is represented on the plane, all the relationships on the spherical surface cannot be duplicated on the plane. Any projection system will involve distortions due to the following points:

(a) Similar angles at different points on the earth or lunar surface *may* or *may not* be shown as similar on the projection system.

(b) The area of a certain region *may* or *may not* be enlarged or reduced relative to the area of another region.

(c) Distance relationships between all points on the earth (or the moon) *cannot* be shown without distortion on the projection plane surface.

(d) Directions between all points on the datum surface *cannot* be shown without distortions on the projection surface.

### 1.5.4 *Selecting or devising a representation system*

The selection of a representation system from the known ones or devising a new system for a given purpose requires first of all the knowledge of the datum surface and the projection surface. The requirements for a desired representation may or may not be possible to satisfy either in part or as a whole. It is, however, possible to classify the requirements and offer the best projection system for the stated purpose from the existing non-conventional ones, often with several alternatives available, expressing their advantages and limitations. Another possibility would be to try to devise a conventional projection, but this would in most cases be a task demanding a very extensive knowledge of mathematics besides the fact that a conventional projection would have an application for the one specific purpose and the connection between this and other systems would be extremely difficult. It is advisable, therefore, to select the best available non-conventional projection, considering possible compromises regarding the original requirements in order to minimize distortions.

## CHAPTER 2

DEFINITIONS, NOTATION, DIMENSIONS

## 2.1  Definitions and notation

### 2.1.1  *Mathematical representation of the shape of the earth and the moon*

The physical shape of the earth and the moon is too irregular to be used directly for computational purposes. Therefore various fictitious mathematical surfaces have been defined approximating the shape to various degrees of precision.

For the purpose of this book, the following review is considered to be sufficient.

The universally accepted best approximation of the shape of the earth is the equipotential surface at mean sea level, called the geoid. It is undulatory, smooth and continuous, fictitiously extending under the continents at the same level, and by definition perpendicular at any point to the direction of gravity. The surface is not symmetrical about the axis of rotation, the distribution of the density within the earth's body being irregular (Fig. 2.1).

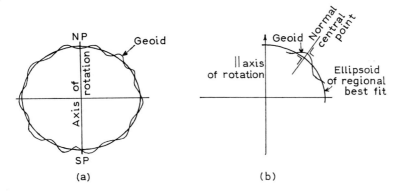

Fig. 2.1
Geoid and ellipsoid

Although the geoid is a convenient medium to study the field of gravity, it has disadvantages if horizontal coordinates of points, distances and angles are to be calculated, because the relevant mathematical formulae become unmanageable. Only heights can be referred conveniently to the geoidal surface but these will not be of concern within the scope of this book.

Another disadvantage is, that the geoid is not completely known since the knowledge of the gravity field of large areas on earth is incomplete. This will of course be remedied in time.

For these reasons a symmetrical surface of revolution is introduced as a best fit of the total geoid. This surface is an ellipsoid the dimensions of which depend on the postulated conditions of best fit (Fig. 2.1a). These are of various charecter and will not be treated here. One common property is that the axis of rotation and the centre are *coincident* respectively with the rotational axis and the centre of gravity of the geoid.

However, as long as the relative position of continents on earth is not known to a sufficient degree of precision it is not possible to map the *whole* world on this ellipsoid. For a partial region of the geoid one introduces an ellipsoid whose surface constitutes a regional best fit only (fig. 2.1b), and whose axis of rotation is *parallel* to that of the geoid. Physical surface points are referred to the ellipsoid by projection. The discrepancies between distances and angles determined on a suitably selected ellipsoid and those on the actual surface are thus kept within limits acceptable for the most precise determination of reference points.

In the process of mapping – representing the curved physical surface through the ellipsoid onto a plane surface – additional discrepancies are being introduced. A large part of these are comprised in the so called "zero" dimension of the map determined by the ultimate precision of plotting. This can be taken as approximately 0.15 mm independent of the scale. The zero dimension at a scale 1:500 then becomes 7.5 cm, it becomes 15 m at a scale 1:100,000. Measurements taken from a map cannot be more precise than its zero dimension.

As long as the influence of the computational differences of the replacement of the ellipsoidal formulae by the simpler spherical ones (taking into account the accuracy of the field measurements) is negligible against the zero dimension of a map, the sphere may take over as an approximation of the physical global figure. Between the outer limits of reference points coordinated on an ellipsoid, the spherical surface is selected so as to be tangent to the ellipsoidal surface at the central point of the region (Fig. 2.2). The radius is a certain function of the parameters of the ellipsoid (see Chapters 5 and 6). The larger the region the larger the deviations of the two surfaces. For mapping the total globe or hemispheres at extremely small scales (atlases) the ellipsoid is completely replaced by a sphere of an appropriate radius.

### 2.1.2  *Definitions and notation on the ellipsoid and the sphere*

For the ensuing chapters it is necessary to define the elements of the ellipsoid as well as the corresponding ones of the sphere. Refer to Figure 2.3 where a part of an ellipsoid is shown. The $z$-axis is the axis of revolution. A plane through this axis intersects the surface along a meridional ellipse. The centre is $O$; the North and South Pole are indicated by NP and SP respectively. The relationships between the elements are given without further proof.

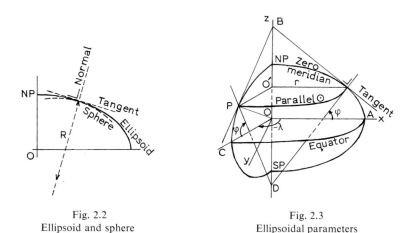

Fig. 2.2
Ellipsoid and sphere

Fig. 2.3
Ellipsoidal parameters

$OA = OC = a =$ the semi major axis of the meridional ellipse = the equatorial radius;

$ONP = b =$ the semi minor axis of the meridional ellipse;

The flattening or compression*

$$f = \frac{a-b}{a}. \tag{2.1}$$

The first eccentricity

$$\varepsilon = \sqrt{\frac{a^2-b^2}{a^2}} \tag{2.2}$$

with

$$\varepsilon^2 = 1-(1-f)^2. \tag{2.3}$$

---

* In the literature also indicated by "$c$".

The second eccentricity

$$\varepsilon' = \sqrt{\frac{a^2 - b^2}{b^2}} \qquad (2.4)$$

with

$$(\varepsilon')^2 = \frac{\varepsilon^2}{1 - \varepsilon^2}. \qquad (2.5)$$

The angle between the normal $PD$ at $P$ and the equatorial axis $CO$ is called the *geodetic latitude* $\varphi$.

The angle $COP$ is the *geocentric* latitude.

The radius of curvature of the meridian at $P$ is $M$; the radius of curvature in the direction perpendicular to the direction of meridian is $N = PD$.

$$M = \frac{a(1 - \varepsilon^2)}{(1 - \varepsilon^2 \sin^2 \varphi)^{3/2}}. \qquad (2.6)$$

The radius of the parallel circle of $P$ ($\varphi$ is constant) is $PO'$ and is equal to

$$PO' = N \cos \varphi = r. \qquad (2.7)$$

The length $PB$ of the tangent at $P$ is equal to

$$PB = N \cot \varphi. \qquad (2.8)$$

The meridian of $P$ is indicated by the angle $\lambda$, called the longitude of $P$, and counted from a reference or zero meridian. This is as a rule the meridian of Greenwich. However, for the treatment of map projections it may be more convenient to count from the meridian of the central point of the region to be mapped. The angle $\lambda$ is usually taken positive Eastward, and negative Westward of this zero meridian.

The radius of curvature $N$ in the direction of the prime vertical perpendicular to the direction of the meridian

$$N = \frac{a}{(1 - \varepsilon^2 \sin^2 \varphi)^{1/2}} \qquad (2.9)$$

The equation of the ellipsoid in a rectangular 3-dimensional coordinate system (as shown in Figure 2.3) is given by

$$\frac{x^2 + y^2}{a^2} + \frac{z^2}{b^2} = 1, \qquad (2.10)$$

or in polar coordinates

$$
\left.\begin{array}{l}
x = N \cos \varphi \cos \lambda \\
y = N \cos \varphi \sin \lambda \\
z = N(1 - \varepsilon^2) \sin \varphi .
\end{array}\right\} \tag{2.11}
$$

The azimuth of an arbitrary arc $PP'$ on the surface is denoted by $\alpha$, counted clockwise from the North direction. This is shown in Figure 2.4.

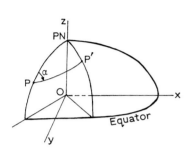

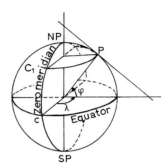

| Fig. 2.4 | Fig. 2.5 |
|---|---|
| Azimuth of $PP'$ | Spherical parameters |

The corresponding formulae for the sphere are obtained by putting $\varepsilon = 0$ in the above expressions (see Fig. 2.5).

There is only one radius of curvature

$$
M = N = R
$$

and of course the geodetic latitude $\varphi$ is coincident with the geocentric latitude.

In the case of the ellipsoid as well as of the sphere the position of points on the surface is indicated by the latitude $\varphi$ counted positive from the equator and the longitude $\lambda$, both expressed in degrees, minutes and seconds of arc.

## 2.2   Lines of special properties

There are a number of lines of special properties on the surface of the ellipsoid (and their spherical equivalents). They are the normal sections, the geodesic and the rhumbline or loxodrome. The normal sections, however, though important in geodesy, are not of interest in map projections.

### 2.2.1   *The geodesic, the great circle, the orthodrome*

The shortest possible connection between $P_1$ and $P_2$ on the ellipsoidal surface is the geodetic line or for convenience "the geodesic". Progressing along this (curved) line from point to point the tangent is continuously changing its azimuth.

The following relationship exists at every point $P$ of the geodesic according to a theorem of Clairaut (see e.g. [29]). "The product of the radius of the parallel circle of $P$ and the sine of the azimuth of the geodesic is a constant", or

$$r \sin \alpha = N \cos \varphi \sin \alpha = C . \tag{2.12}$$

Some particulars can be derived from this formula.

(1)  For $\varphi = 0$:  $N = a$ and

$$\sin \alpha = \frac{C}{a} . \tag{2.13}$$

The geodesic intersects the equator, the azimuth $\alpha$ being equal to arc $\sin \dfrac{C}{a}$

(2)  For  $\alpha = 90°$,
     it is seen that

$$N \cos \varphi = C \Big]$$

and

$$N \cos - \varphi = C \Big] \tag{2.14}$$

respectively. The solution of these equations yields two values of $\varphi$ symmetrical about the equator.

(3)  For $\alpha = 0$ the $\varphi$ is undetermined indicating a meridian:
     The meridian is a geodesic.

An illustration is shown in Figure 2.6.

If $P_1$ and $P_2$ have the maximum difference in longitude $\lambda_2 - \lambda_1 = 180°$ the shortest way is along a meridional section through one of the poles.

The parallel circles are not geodesic lines, although the equation (2.12) is satisfied.

Both the normal sections between two points as well as the geodesic merge into part of a great circle on the sphere since it is a particular case of the ellipsoid when the semi major and semi minor axes become equal. Hence the shortest connection between two points on the sphere is part of a great circle. This curve, however, is not portrayed as a straight line in any but a central or gnomonic projection. The projected curve is sometimes called the "ortho-drome". It is of great importance in "great circle" navigation, and for finding direction by wireless.

The orthodrome can be plotted on a map in two ways, viz.:

(1) By calculating the latitude and longitude of various points of the line progressing from the starting point $P_1$ to the terminal $P_2$ by spherical trigonometry; plotting them on the map, and connecting them with a continuous curve.

The formulae are derived from the polar triangle $P_1 NP P_2$ (see Fig. 2.7 where

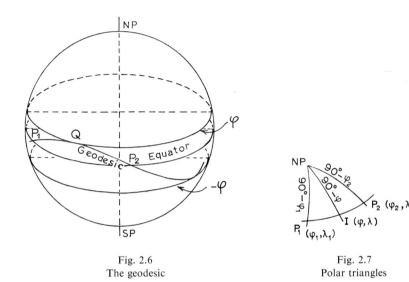

Fig. 2.6
The geodesic

Fig. 2.7
Polar triangles

one intermediate point $I$ is shown), and are suitable for programming. Counting $\lambda$ positive in the East direction it is seen that

$$\tan \varphi = \frac{\cot \alpha \sin (\lambda - \lambda_1) + \sin \varphi_1 \cos (\lambda - \lambda_1)}{\cos \varphi_1}. \qquad (2.15)$$

Since

$$\cot \alpha = \frac{\cos \varphi_1 \tan \varphi_2 - \sin \varphi_1 \cos (\lambda_2 - \lambda_1)}{\sin (\lambda_2 - \lambda_1)}, \qquad (2.16)$$

it is seen that after some manipulation

$$\left. \tan \varphi = \frac{\tan \varphi_2 \cos \lambda_1 - \tan \varphi_1 \cos \lambda_2}{\sin (\lambda_2 - \lambda_1)} \sin \lambda - \\ - \frac{\tan \varphi_2 \sin \lambda_1 - \tan \varphi_1 \sin \lambda_2}{\sin (\lambda_2 - \lambda_1)} \cos \lambda. \right\} \qquad (2.17)$$

The coefficients of $\sin \alpha$ and $\cos \alpha$ are constants for the calculation of $\varphi$ for any longitude $\lambda$ with $\lambda_2 > \lambda > \lambda_1$.

(2) By drawing a linear connection between $P_1$ and $P_2$ in a map of gnomonic projection. The latitude and longitude of the intersection points with the meridian and parallels can then be transferred to a map of any other non gnomonic projection thus generating the orthodromic curve.

The length of the curve is equal to

$$d_{P_1 P_2} = R \frac{\beta}{\rho}, \qquad (2.18)$$

where R is the radius of the sphere;

$\beta$ the central angle subtended by the arc of the orthodrome $P_1 P_2$ in degrees;

$\rho = 57°29578$.

The angle $\beta$ is calculated by the cosine rule in $\Delta P_1 NP P_2$

$$\cos \beta = \sin \varphi_1 \sin \varphi_2 + \cos \varphi_1 \cos \varphi_2 \cos(\lambda_2 - \lambda_1). \qquad (2.19)$$

### 2.2.2   The rhumbline or loxodrome

The rhumbline between two points $P_1$ and $P_2$ is a curve intersecting the meridians at a constant azimuth $\alpha$ (Fig. 2.8). The equation of the curve is generated from the differential equation

$$d\lambda = \tan \alpha \cdot \frac{1}{\cos \varphi} d\varphi \qquad (2.20)$$

derived directly from Figure 2.9.

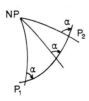

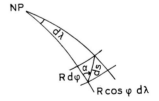

Fig. 2.8                   Fig. 2.9

Rhumbline          Differential elements rhumbline

By integration to point $P_2$ and solving for $\tan \alpha$,

$$\tan \alpha = \frac{\lambda_2 - \lambda_1}{\ln \tan (45° + \tfrac{1}{2}\varphi_2) - \ln \tan (45° + \tfrac{1}{2}\varphi_1)}. \qquad (2.21)$$

Any other point $P(\varphi, \lambda)$ of the curve can now be calculated by integration of (2.20) with boundaries $\lambda_1$ and $\lambda$:

$$\lambda - \lambda_1 = \tan\alpha \left\{\ln\tan(45° + \tfrac{1}{2}\varphi) - \ln\tan(45° + \tfrac{1}{2}\varphi_1)\right\}. \tag{2.22}$$

Particular features of the curve are

(1) If $\varphi_1 = \varphi_2$, $\tan\alpha = \infty$ or $\alpha = 90°$, rendering a parallel circle.

(2) If $\lambda_1 = \lambda_2$, $\alpha = 0$, yielding a meridian.

(3) If $\varphi = 90°$, $\lambda - \lambda_1$ becomes infinite. This means that the curve spirals towards the pole as an asymptotic point (see Fig. 2.10).

The distance $P_1 P_2$ along the loxodrome is equal to

$$P_1 P_2 = \int_{P_1}^{P_2} ds = \frac{R}{\cos\alpha}\,d\varphi = \frac{R}{\cos\alpha}(\varphi_2 - \varphi_1). \tag{2.23}$$

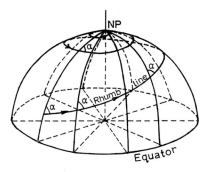

Fig. 2.10
Rhumbline or loxodrome

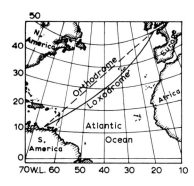

Fig. 2.11
Orthodrome and loxodrome in
gnomonic projection, $\varphi_0 = 30°$

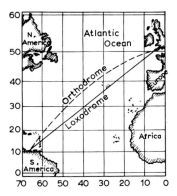

Fig. 2.12
Orthodrome and loxodrome in
Mercator projection

The rhumbline is of importance in navigation particularly since it is projected as a straight line between points in the conformal cylindrical projection of Mercator. The expression $\ln \tan(45° + \frac{1}{2}\varphi)$ will frequently occur in Chapter 5, treating the conformal projections. Figures 2.11 and 2.12 show the orthodrome and the loxodrome in a gnomonic and in the conformal Mercator projection respectively (after [30]).

## 2.3  Dimensions

### 2.3.1  *Dimensions of the terrestrial and lunar ellipsoids*

The actual dimensions of the terrestrial ellipsoid have been calculated many times, depending on additional information becoming available. Well-known names connected with these calculations are Airy, Bessel, Clarke, Hayford, Helmert and Krassovsky. Hayford published dimensions of an ellipsoid of best fit for the whole of North America. These have been adopted by the International Union of Geodesy and Geophysics in 1924 as *the* International Ellipsoid of best fit for the total geoid. However, not all countries recalculated coordinates on this surface. In 1967 the International Union adopted new values, where all modern data including those from satellites have been incorporated. A review – both historical and mathematical – of ellipsoidal dimensions has been published in [25].

It is important that all parts of the earth be calculated on one single ellipsoid, especially in view of the mass of global data now becoming available, and of the continuous demand for a higher precision for the purpose of geophysical studies and navigation.

While using maps, one must be aware of the kind of regional ellipsoid that has been adopted. Discrepancies of 400 ft in positions on the North Sea have been reported to be caused only by the application of different ellipsoids for maps of different nations. Intercontinental discrepancies may be many times larger.

The dimensions of various ellipsoids are tabulated in Table 2.1.

The most refined modern measurements have indicated that the equatorial section of both the earth and the moon are not circular but elliptic. The differences of the axes, however, are too small to be of practical significacne for map projections.

The parameters of the lunar ellipsoid according to the Apollo satellite program values are (see also Section 7.6.2)

$a_{\text{max}} = 1738.57$ km, aligned to the earth–moon line

$a_{\text{min}} = 1738.21$ km

$b \quad = 1737.49$ km.

TABLE 2.1

*Terrestrial ellipsoidal dimensions*

| Year Name | Equatorial axis $a$ (metres) | Polar axis $b$ (metres) | Compression 1: | User (not exhaustive) |
|---|---|---|---|---|
| 1830 Everest | 6377276 | 6356075 | 300.8 | India |
| 1841 Bessel | 6377397 | 6356079 | 299.15 | Germany; Indonesia; The Netherlands |
| 1858 Airy | 6377563 | 6356257 | 299.33 | Great Britain |
| 1858 Clarke | 6378294 | 6356619 | 294.3 | |
| 1866 Clarke | 6378206 | 6356584 | 295.0 | United States of America |
| 1880 Clarke | 6378249 | 6356515 | 293.5 | South Africa |
| 1909 Hayford | 6378388 | 6356912 | 297.0 | United States of America, adopted internationally 1924 |
| 1948 Krassovsky | 6378245 | 6356863 | 298.3 | Russia, Eastern countries |
| 1967 I.U.G.G. | 6378160 | 6356775 | 298.25 | Adopted internationally |

### 2.3.2   *Spherical and plane surfaces as approximations*

It is an important feature for both geographers and geodesists to know the extent of the ellipsoidal surface which can be considered spherical, and similarly to what extent the ellipsoidal and spherical surfaces can be considered plane. What is the small area of the tangent sphere or plane where spherical or plane trigonometry respectively can be used ? The problem is of greater importance for conformal than for equivalent projections, the latter generally being applied at smaller scales and less concerned with the computation of angles and distances on the physical surface. One may therefore formulate a difference for various areas as a tolerance, and decide whether this is or is not of influence on the intended use of a map. Driencourt [7] solved the problem of the approximation of the ellipsoid by a sphere tangent at the central point, onto which angles are transferred conformally. The radius $R$ of this sphere is equal to the geometric mean of the two radii of curvature at the central point.

$$R = \sqrt{M_0 N_0}. \tag{2.24}$$

This is derived in Section 5.2.

A curve is described on the ellipsoid by rotating a geodesic of a constant length $s$ about the central point $O$ $(\varphi_0 \lambda_0)$ through $360°$. This curve is called a geodetic circle. The angle of rotation is $\vartheta$ counted clockwise from the meridian of $O$ (Fig. 2.13). The corresponding situation on the sphere is shown in Figure 2.14. The geodesic has become part of a great circle of length $s$. The angle of rotation is also equal to $\vartheta$.

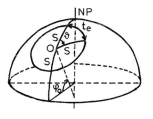 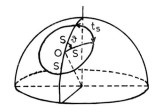

Fig. 2.13
Ellipsoid

Fig. 2.14
Sphere tangent to ellipsoid at $O$

The difference of the arc lengths of the part $t_e$ $(AB)$ of the geodetic circle, and its spherical equivalent $t_s$ is equal to

$$t_e - t_s = \frac{s^4 \varepsilon^2 \sin \varphi_0 \cos \varphi_0}{3 N_0^3 (1 - \varepsilon^2)} \sin \vartheta \tag{2.25}$$

apart from terms of the fifth order and higher. For the derivation of this formula refer to [7].

The difference $t_e - t_s$ is equal to zero for $\varphi = 0$ and $\varphi = 180°$. It fluctuates between

$$(t_e - t_s)_{max} = \frac{s^4 \varepsilon^2 \sin \varphi_0 \cos \varphi_0}{3 N_0^3 (1 - \varepsilon^2)} \qquad \text{for } \vartheta = 90°$$

and

$$(t_e - t_s)_{min} = -\frac{s^4 \varepsilon^2 \sin \varphi_0 \cos \varphi_0}{3 N_0^3 (1 - \varepsilon^2)} \qquad \text{for } \vartheta = 270°. \tag{2.26}$$

If the ellipsoidal surface is pressed upon the sphere (keeping the meridional arcs through $O$ on both surfaces in coincidence) the discrepancy between the circumferences of the geodetic and the spherical circles becomes evident as an angular split starting at $O$. The total discrepancy is $2(t_e - t_s) = \Delta t$ at a radial distance $s$ (see Fig. 2.15). For $\varphi_0 = 0$ or $\varphi_0 = 90°$ i.e. with the central point $O$ on the equator or in the pole respectively the $t_e - t_s = 0$.

Fig. 2.15
Discrepancy $\Delta t$

An absolute maximum $\Delta t$ is obtained for $\varphi_0 = 45°$

$$\Delta t = \frac{2s^4}{3a^3} \cdot \frac{\varepsilon^2 (1 - \tfrac{1}{2}\varepsilon^2)^{3/2}}{(1 - \varepsilon^2)}. \tag{2.27}$$

Driencourt tabulated the $\Delta t$ for various radial distances $s$ on the terrestrial ellipsoid (Table 2.2).

TABLE 2.2

*Discrepancy $\Delta t$ (terrestrial)*

| s (km) | 103.7 | 184.4 | 327.9 | 583.1 | 1039.9 |
|---|---|---|---|---|---|
| $\Delta t$ (m) | 0.001 | 0.01 | 0.10 | 1.00 | 10.00 |
| Angular (radians) | 0.016 | 0.029 | 0.051 | 0.09 | 0.16 |

This table shows that the discrepancy $\Delta t$ at a distance of approximately 100 km from the central point, does not exceed 1 mm or $10^{-8}s$. At a distance of 1000 km from $O$ the proportion $\Delta t/s$ is approximately $10^{-5}$ which is about three times the present precision of electronic distance measurement.

The lunar surface (with $a_{mean} = 1738.39$ km and $\varepsilon^2 = 0.001035$) shows the values as represented in Table 2.3.

The conversion into radians (see Tables 2.2 and 2.3 third line) shows clearly the lunar ellipsoid being "more spherical" than the terrestrial ellipsoid. If $\Delta t$

TABLE 2.3

*Discrepancy $\Delta t$ (lunar)*

| $s$ (km) | 63 | 112 | 199 | 354 | 630 |
|---|---|---|---|---|---|
| $\Delta t$ (m) | 0.001 | 0.01 | 0.10 | 1.00 | 10.00 |
| Angular (radians) | 0.036 | 0.064 | 0.114 | 0.20 | 0.36 |

is equal for instance to 10 cm, an area of approximately $3.1 \times 10^6$ km² or 0.06% of the total surface of the earth can be replaced by a sphere, whilst on the moon approximately $1.2 \times 10^6$ km² or 0.3% of the total surface area can be replaced.

The discrepancies occurring by laying out a part of a spherical surface upon a plane are judged in the following way [7]. In Figure 2.16 a spherical segment having a (circular) circumference $ABCA$ is assumed to consist of an infinite number of spherical isosceles triangles with the apex at $O$. They are replaced by plane triangles (Fig. 2.17) according to Legendre's rule which states that the sides of a spherical triangle can be computed by the trigonometry of a plane triangle the sides of which have the same length as those of the spherical triangle, the plane angles being equal to the corresponding spherical angles diminished by one third of the spherical excess $e_x$.

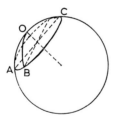

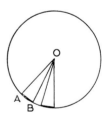

Fig. 2.16
Spherical triangle

Fig. 2.17
Plane triangle

The spherical excess of a spherical triangle is equal to

$$e_x = \frac{\text{Area}}{R^2} \rho'', \tag{2.28}$$

where Area is the area of the triangle,

    R the radius of the sphere,

    $\rho'' = 206264''.8$

Hence the sum of the plane angles at $O$ (Fig. 2.17) will be equal to

$$360° - \frac{\text{Area of the segment}}{3R^2}\rho''. \qquad (2.29)$$

Laying out the plane triangles contiguous with their apices in one central point, the term $\frac{1}{3}e_x$ becomes a measure for the discrepancy between the spherical surface and the tangent plane at $O$.

Driencourt also tabulated the corresponding linear values of $\frac{1}{3}e_x$ on earth [7] for various ranges of $OA$ (in kilometres) about the central point. These linear values are given in metres (Table 2.4).

TABLE 2.4
*Spherical discrepancies (terrestrial)*

| $OA$ (km) | $\frac{1}{3}e_x$ | Corr. linear (m) | Approx. Area km² |
|---|---|---|---|
| 7.3 | 0″.3 | 0.01 | 167 |
| 15.7 | 1″.3 | 0.10 | 772 |
| 33.9 | 6″.1 | 1.00 | 3630 |
| 72.9 | 28″.31 | 10.00 | 16730 |
| 100.0 | 53″.26 | 25.74 | 31400 |

The corresponding figures for the lunar surface are given in Table 2.5.

TABLE 2.5
*Spherical discrepancies (lunar)*

| $OA$ (km) | $\frac{1}{3}e_x$ | Corr. linear (m) | Approx. Area km² |
|---|---|---|---|
| 3.1 | 0″.68 | 0.01 | 30 |
| 6.7 | 3″.2 | 0.10 | 140 |
| 14.4 | 14″.6 | 1.00 | 650 |
| 31.0 | 68″.5 | 10.00 | 3020 |
| 50.0 | 178″.4 | 43.20 | 7850 |

By considering the tolerable discrepancy in a particular case one can decide upon the area to be considered plane.

## 2.4  Some remarks on coordinate systems

### 2.4.1  *Origin and grid*

One must adopt some coordinate system in the projection plane in order to be able to plot the relative position of ellipsoidal or spherical coordinates $(\varphi, \lambda)$ of meridians, parallel circles and other features to be projected. For this it is necessary:

(1) to decide upon the nature of this plane system;

(2) to select the position of an origin;

(3) to relate the orientation of the ellipsoidal and plane systems;

(4) to select suitable units of measure along the coordinate axes.

For the sake of simplicity, the normal projection is considered first (Section 1.3.1).

Commonly a rectangular Cartesian coordinate system is adopted with a positive $X$-axis serving as an abscissa pointing North. This axis is sometimes called "Northing". The positive $Y$-axis points East ("Easting"). Accordingly, the negative $X$ may be denoted as "Southing" and the negative $Y$-axis as "Westing". The system is called a grid. Sometimes it is convenient to apply polar coordinates as a transistory system between the $(\varphi, \lambda)$ and the $XY$ systems in order to facilitate the computations.

The origin may be selected so as to correspond with the central point of the projected area, where the ellipsoidal normal coincides with the normal to the local geoid (the origin of the regional geodetic computations). In the case of conical or cylindric projections this central point may be situated on the tangent parallel or the meridian. This selection is convenient but not strictly necessary.

The relationship with the ellipsoidal or spherical coordinate system is determined by the type of projection and the pertinent formulae.

$$\left. \begin{array}{l} X = f_1(\varphi, \lambda) \\ Y = f_2(\varphi, \lambda) \, . \end{array} \right\} \tag{2.30}$$

These expressions fix the orientation of one system relative to the other. In the origin $O$ with coordinates $(\varphi_0, \lambda_0)$ (2.30) becomes

$$\left. \begin{array}{l} f_1(\varphi_0, \lambda_0) = 0 \\ f_2(\varphi_0, \lambda_0) = 0 \, . \end{array} \right\} \tag{2.31}$$

The choice of the units of measurement depends largely on the purpose of the map. They can be feet, miles, chains (!), meters etc. Unless stated otherwise the metric system will be used.

In the ensuing derivations it will be assumed, that the earth and the moon are depicted in their *natural* dimensions with the scale distortion or magnification* in the origin equal to unity. Deviations from this unit factor away from the origin are inherent to the particular type of projection. In Chapter 5 it is treated in what manner in conformal projections the distribution of these deviations from the unit value may be improved by the application of a scale factor ($<1$) at the origin. It will at present be of no concern at what scale the image is viewed on the map.

By programming the formulae (2.30) (in combination with a mechanical plotting device) or by means of existing tabulations the meridians and parallels are plotted in the $X$, $Y$ system, thus forming a plane network (Fig. 2.18). The necessary density depends also on the purpose of the map. In relation to the scale it may be densified until a linear interpolation is possible. In Figure 2.18 $AB$ is the central meridian which one usually wants to be projected as a straight line. The straight line connection $PQ$ forms an angle $\hat{A}$ about the North direction (Fig. 2.19); it is called the grid bearing of $PQ$ and counted clockwise through 180° to 360°.

Figure 2.20 shows a polar region.

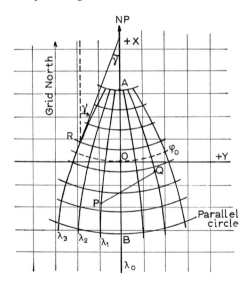

Fig. 2.18
Meridians and parallels in a plane $X,Y$ grid

At $R$ in Figure 2.18 the angle $\gamma$ between the $+X$ direction (grid North) and the direction of the tangent to the meridian is called the meridian convergence

---

* Often called the "projection scale factor".

at $R$. Where meridians are projected as straight lines the convergence is formed by the $+X$ direction and the meridian itself. (See $\not\!\star\gamma$ at point $R$ in Figure 2.20). It is a variable quantity counted positive and negative, with a possible maximum of $180°$.

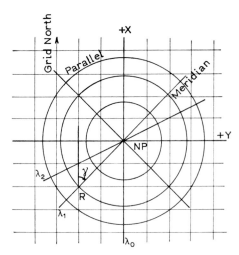

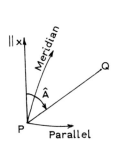

Fig. 2.19
Grid bearing

Fig. 2.20
Meridians and parallels in a $X$, $Y$ grid (polar region)

A mathematical expression is derived in Appendix A. It appears that at first approximation the convergence is *independent* of the projection, viz.

$$\gamma = \lambda \sin \varphi \qquad\qquad (2.32)$$

It suffices for any but very precise purposes to calculate $\gamma$ with this formula.

It may be felt as a practical difficulty that the $X$, $Y$ coordinates become unduly large or are of different signs in the four quadrants. In that case a translation of the origin $O$ can be applied, introducing a new origin (known by the name of "false" origin) so as to render all new coordinates positive. In many instances where the projected region is subdivided into parts for the convenience of mapping, each map or plan has its own false origin (Fig. 2.21).

It frequently occurs especially in small scale maps and atlases, that the $XY$ grid does not appear.

### 2.4.2    Transformation of coordinates for oblique projections

For the purpose of introducing oblique projections the earth or moon are considered spherical (it may be assumed that the ellipsoid has been projected to a sphere as an intermediate step to the plane projection).

In the case of the projection surface being in a normal position the relationships between the spherical coordinates, and those in the projection surface $X$, $Y$ can be derived directly. With oblique projections, however, this cannot be done.

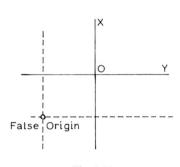

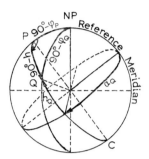

Fig. 2.21
False origin

Fig. 2.22
Oblique projection surface

Suppose the axis of the projection surface coincides (or is parallel) with $PC$ (see Fig. 2.22). Then $P$ acts for an oblique projection in the same manner as the North Pole in a normal projection. The great circle in the plane perpendicular to $PC$ acts as an equator.

Taking the meridian through the North Pole and $P$ as a reference meridian an arbitrary point $Q$ is located in the oblique system by the arcs $\alpha_Q$ and $h_Q$ or the polar coordinates $\alpha_Q$ and $(90-h_Q)$. All projection formulae can be derived directly in this $(h, \alpha)$ system. It is then necessary to transform back to the $(\varphi, \lambda)$ system for further mapping of the meridians and parallels by spherical trigonometry. Knowing that the Pole $P$ has coordinates $(\varphi_P, \lambda_P)$ the latitude of $Q$ is formed in $\Delta P(\mathrm{NP})Q$ as

$$\sin \varphi_Q = \sin \varphi_P \sin h_Q + \cos \varphi_P \cos h_Q \cos \alpha_Q \tag{2.32}$$

and the longitude $\lambda_Q$ by

$$\sin (\lambda_Q - \lambda_P) = \frac{\sin \alpha_Q \cos h_Q}{\cos \varphi_Q} \tag{2.33}$$

In the extreme case if $\varphi_{\mathrm{Pole}} = 0$ a transverse projection is obtained. Also

$$\cot (\lambda_Q - \lambda_P) = \frac{\sin \varphi_P \cos \alpha_Q + \cos \varphi_P \tan h_Q}{\sin \alpha_Q}. \tag{2.34}$$

The inverse formulae are

$$\left.\begin{aligned}
\sin h_Q &= \sin \varphi_Q \sin \varphi_P + \cos \varphi_Q \cos \varphi_P \cos (\lambda_Q - \lambda_P) \\
\cos \alpha_Q \cos h_Q &= \sin \varphi_Q \cos \varphi_P - \cos \varphi_Q \sin \varphi_P \cos (\lambda_Q - \lambda_P) \\
\tan \alpha_Q &= \frac{\sin (\lambda_Q - \lambda_P) \cos \varphi_Q}{\sin \varphi_Q \cos \varphi_P - \cos \varphi_Q \sin \varphi_P \cos (\lambda_Q - \lambda_P)}
\end{aligned}\right\} \quad (2.35)$$

In the projection plane the origin is chosen in the central point, whilst the $+X$ direction is pointed towards the pole $P$ instead of to the North Pole (see Fig. 2.23).

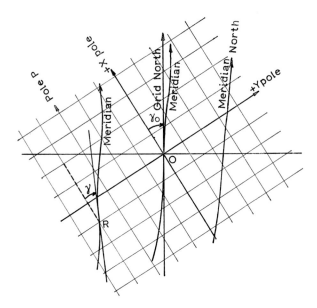

Fig. 2.23
Oblique projection surface

The projection formulae now become

$$\left.\begin{aligned}
X_{\text{Pole}} &= f_1(h, \alpha) \\
Y_{\text{Pole}} &= f_2(h, \alpha)
\end{aligned}\right\} \quad (2.36)$$

By combination of (2.32), (2.33), and (2.36) the meridians and parallels are plotted as curved lines such as schematically indicated in Figure 2.23.

The meridian convergence $\gamma$ at a point $R$ is defined in a manner similar as before viz.: the angle between the $X_{\text{Pole}}$ direction and the tangent to the meridian at that point. It may be called the oblique convergence.

For reasons of convenience a grid $X_{\text{North}}\ Y_{\text{East}}$ is wanted on a map oriented to the true North. This is achieved by rotation of the coordinate system ($X_{\text{Pole}}$, $Y_{\text{Pole}}$) about an angle equal to the meridian convergence in the origin (or for each map separately in the false origin).

The axes only are drawn in Figure 2.23. The rotation formulae are

$$\left.\begin{aligned}
X_N &= X_P \cos \gamma_0 + Y_P \sin \gamma_0 \\
Y_N &= -X_P \sin \gamma_0 + Y_P \cos \gamma_0 \ .
\end{aligned}\right\} \tag{2.37}$$

CHAPTER 3

# GENERAL TRANSFORMATION FORMULAE.
# THEORY OF DISTORTIONS

### 3.1 General transformation formulae; conditions of uniqueness, reversibility and correspondence of parametric curves

As has been shown in previous chapters, a system of reference curves or parametric curves can be adopted on a curved surface. If these curves are denoted as $u$ and $v$ curves respectively then any point on this surface may be given in Cartesian coordinates $x$, $y$, $z$ as functions of $u$ and $v$ (Fig. 3.1).

$$\left. \begin{aligned} x &= p_1(u, v) \\ y &= p_2(u, v) \\ z &= p_3(u, v) \end{aligned} \right\} \tag{3.1}$$

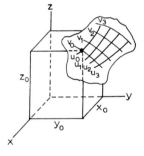 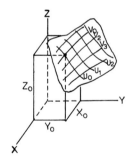

| Fig. 3.1 | Fig. 3.2 |
|:---:|:---:|
| Datum surface | Projection surface |

For convenience this surface will be called the datum surface. The same relationships can be written for a second surface called the projection or image surface (Fig. 3.2).

$$\left. \begin{aligned} X &= p_1(U, V) \\ Y &= p_2(U, V) \\ Z &= p_3(U, V). \end{aligned} \right\} \tag{3.2}$$

34

The parametric curves on the former surface are related to some arbitrary system of curves on the latter surface if there exists a mathematical relationship between the parameters $u$, $v$ and $U$, $V$, viz.

$$\left.\begin{array}{l} U = q_1(u, v) \\ V = q_2(u, v). \end{array}\right\} \tag{3.3}$$

It is obvious that if the surface of the earth or moon is to be depicted on a sphere or in a plane, two conditions should be satisfied:

(1) the projection or image should be *unique*;
(2) the projection should be *reversible*.

This means that a point on the datum surface should correspond to one and only one point on the projection surface. The reverse must also hold.

Mathematically this can be expressed by the condition that the parameters $u$ and $v$ should be solvable from the equations (3.3), yielding them as explicit functions of $U$ and $V$.

$$\left.\begin{array}{l} u = \bar{q}_1(U, V) \\ v = \bar{q}_2(U, V). \end{array}\right\} \tag{3.4}$$

Without further restrictions the parametric curves (3.4) $u$ and $v$ do not as a rule correspond with the system $(U, V)$, but become an arbitrary different system. E.g. the curve $u = 5$ becomes

$$5 = \bar{q}_1(U, V)$$

and not $U = 5$ or $V = 5$.

However they can be made to be so by eliminating the $U$ and $V$ from the equations (3.1) and (3.2) with the aid of (3.3).

This gives the general transformation or projection formulae

$$\left.\begin{array}{ll} x = p_1(u, v) & X = \bar{p}_1(u, v) \\ y = p_2(u, v) \quad \text{and} & Y = \bar{p}_2(u, v) \\ z = p_3(u, v) & Z = \bar{p}_3(u, v). \end{array}\right\} \tag{3.5}$$

## 3.2   Some elementary differential geometry. The transformation matrix

### 3.2.1   *The first Gaussian fundamental quantities. Angular expressions*

According to a formula of the differential geometry the square of the length of an elementary part of a curve on a surface is given by

$$ds^2 = dx^2 + dy^2 + dz^2 \tag{3.6}$$

By differentiating (3.1)

$$
\left.
\begin{aligned}
\mathrm{d}x &= \frac{\partial x}{\partial u}\,\mathrm{d}u + \frac{\partial x}{\partial v}\,\mathrm{d}v \\[1.2em]
\mathrm{d}y &= \frac{\partial y}{\partial u}\,\mathrm{d}u + \frac{\partial y}{\partial v}\,\mathrm{d}v \\[1.2em]
\mathrm{d}z &= \frac{\partial z}{\partial u}\,\mathrm{d}u + \frac{\partial z}{\partial v}\,\mathrm{d}v
\end{aligned}
\right\}
\tag{3.7}
$$

whence by substitution of (3.7) into (3.6)

$$
\left.
\begin{aligned}
\mathrm{d}s^2 = & \left\{ \left(\frac{\partial x}{\partial u}\right)^2 + \left(\frac{\partial y}{\partial u}\right)^2 + \left(\frac{\partial z}{\partial u}\right)^2 \right\} \mathrm{d}u^2 + \\[1em]
& + 2\left\{ \frac{\partial x}{\partial u}\cdot\frac{\partial x}{\partial v} + \frac{\partial y}{\partial u}\cdot\frac{\partial y}{\partial v} + \frac{\partial z}{\partial u}\cdot\frac{\partial z}{\partial v} \right\} \mathrm{d}u\,\mathrm{d}v + \\[1em]
& + \left\{ \left(\frac{\partial x}{\partial v}\right)^2 + \left(\frac{\partial y}{\partial v}\right)^2 + \left(\frac{\partial z}{\partial v}\right)^2 \right\} \mathrm{d}v^2
\end{aligned}
\right\}
\tag{3.8}
$$

The coefficients of $\mathrm{d}u^2$, $\mathrm{d}u\,\mathrm{d}v$ and $\mathrm{d}v^2$ are called *the first Gaussian fundamental quantities*, and are invariably indicated in the literature by $E$, $F$, $G$ (or $e$, $f$, $g$) respectively.

Hence (3.6) becomes

$$
\mathrm{d}s^2 = \mathrm{e}\,\mathrm{d}u^2 + 2\mathrm{f}\,\mathrm{d}u\,\mathrm{d}v + \mathrm{g}\,\mathrm{d}v^2
\tag{3.9}
$$

The quantities $\sqrt{e}$ and $\sqrt{g}$ act as units of measure along the $u$ and $v$ curves on the surface. In the Figure 3.3 the differential parallelogram is shown at a point $P$ of a curve.

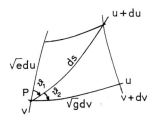

Fig. 3.3
Differential parallelogram

Now

$$\vartheta_1 + \vartheta_2 = \omega, \tag{3.10}$$

where $\omega$ is the angle of intersection of the $u$ and $v$ curves through $P$. The parallelogram may be considered as plane within its infinite small area, so that by application of the cosine rule

$$ds^2 = e \, du^2 + g \, dv^2 + 2\sqrt{eg} \, du \, dv \cos \omega. \tag{3.11}$$

Equating (3.11) and (3.9) gives

$$\cos \omega = \frac{f}{\sqrt{eg}}. \tag{3.12}$$

Further

$$\cos \vartheta_1 = \frac{\sqrt{e} \, du + \sqrt{g} \, dv \cos \omega}{ds} = \frac{1}{\sqrt{e}}\left(e \frac{du}{ds} + f \frac{dv}{ds}\right) \tag{3.13}$$

and

$$\cos \vartheta_2 = \frac{\sqrt{g} \, dv + \sqrt{e} \, du \cos \omega}{ds} = \frac{1}{\sqrt{g}}\left(f \frac{du}{ds} + g \frac{dv}{ds}\right). \tag{3.14}$$

Also

$$\sin \omega = \frac{\sqrt{eg - f^2}}{\sqrt{eg}}; \tag{3.15}$$

$$\sin \vartheta_1 = \sqrt{g} \sin \omega \frac{dv}{ds} = \frac{\sqrt{eg - f^2}}{\sqrt{e}} \frac{dv}{ds} \tag{3.16}$$

$$\sin \vartheta_2 = \sqrt{e} \sin \omega \frac{du}{ds} = \frac{\sqrt{eg - f^2}}{\sqrt{g}} \frac{du}{ds}. \tag{3.17}$$

The angle $\vartheta_1$ will be denoted as the bearing of $ds$.
The elementary area

$$A_D = \sqrt{eg} \sin \omega \, du \, dv = \sqrt{eg - f^2} \, du \, dv. \tag{3.18}$$

The expression $eg - f^2$ is definite positive; it occurs frequently in the following derivations.

The same formulae with capital $E$, $F$, $G$, $\Omega$, $\Theta$ are applicable to the projection surfaces.

In an orthogonal parametric system $(u, v)$ where $\omega = 90°$, the quantity $f = 0$.

### 3.2.2    *The fundamental transformation matrix; the Jacobean determinant*

Reference is made to the formulae (3.1) to (3.5) inclusive. The fundamental quantities pertaining to (3.1) are indicated by $e$, $f$ and $g$; to (3.2) by $E'$, $F'$, $G'$; and to the projection surface of (3.5) by $E$, $F$, $G$.

Differentiating (3.3) and (3.5) gives

$$\left.\begin{aligned}
\frac{\partial X}{\partial u} &= \frac{\partial X}{\partial U}\cdot\frac{\partial U}{\partial u} + \frac{\partial X}{\partial V}\frac{\partial V}{\partial u} \\[2mm]
\frac{\partial X}{\partial v} &= \frac{\partial X}{\partial U}\frac{\partial U}{\partial v} + \frac{\partial X}{\partial V}\frac{\partial V}{\partial v}\,.
\end{aligned}\right\} \tag{3.19}$$

Similarly

$$\frac{\partial Y}{\partial u}\,;\;\frac{\partial Y}{\partial v}\,;\;\frac{\partial Z}{\partial u}\,;\;\frac{\partial Z}{\partial v}\,;$$

are obtained.

Hence after some derivations

$$\left.\begin{aligned}
E &= \left(\frac{\partial X}{\partial u}\right)^2 + \left(\frac{\partial Y}{\partial u}\right)^2 + \left(\frac{\partial Z}{\partial u}\right)^2 = \left(\frac{\partial U}{\partial u}\right)^2 E' + 2\frac{\partial U}{\partial u}\frac{\partial V}{\partial u}F' + \left(\frac{\partial V}{\partial v}\right)^2 G' \\[2mm]
F &= \frac{\partial X}{\partial u}\frac{\partial X}{\partial v} + \frac{\partial Y}{\partial u}\frac{\partial Y}{\partial v} + \frac{\partial Z}{\partial u}\frac{\partial Z}{\partial v} = \\[2mm]
&= \left(\frac{\partial U}{\partial u}\frac{\partial U}{\partial v}\right)E' + \left(\frac{\partial U}{\partial u}\frac{\partial V}{\partial v} + \frac{\partial U}{\partial v}\frac{\partial V}{\partial u}\right)F' + \left(\frac{\partial V}{\partial u}\frac{\partial V}{\partial v}\right)G' \\[2mm]
G &= \left(\frac{\partial X}{\partial v}\right)^2 + \left(\frac{\partial Y}{\partial v}\right)^2 + \left(\frac{\partial Z}{\partial v}\right)^2 = \left(\frac{\partial U}{\partial v}\right)^2 E' + 2\frac{\partial U}{\partial v}\frac{\partial V}{\partial v}F' + \left(\frac{\partial V}{\partial v}\right)^2 G',
\end{aligned}\right\} \tag{3.20}$$

or in matrix notation:

$$\begin{bmatrix} E \\[4mm] F \\[4mm] G \end{bmatrix} = \begin{bmatrix} \left(\dfrac{\partial U}{\partial u}\right)^2 & 2\dfrac{\partial U}{\partial u}\dfrac{\partial V}{\partial u} & \left(\dfrac{\partial V}{\partial u}\right)^2 \\[4mm] \dfrac{\partial U}{\partial u}\dfrac{\partial U}{\partial v} & \dfrac{\partial U}{\partial u}\dfrac{\partial V}{\partial v} + \dfrac{\partial U}{\partial v}\dfrac{\partial V}{\partial u} & \dfrac{\partial V}{\partial u}\dfrac{\partial V}{\partial v} \\[4mm] \left(\dfrac{\partial U}{\partial v}\right)^2 & 2\dfrac{\partial U}{\partial v}\dfrac{\partial V}{\partial v} & \left(\dfrac{\partial V}{\partial v}\right)^2 \end{bmatrix} \begin{bmatrix} E' \\[4mm] F' \\[4mm] G' \end{bmatrix}. \tag{3.21}$$

These equations are extremely important for the design of computer pro-grammes. The matrix $[\partial U/\partial u)^2$ etc] will be called the *fundamental transformation matrix*.

The term $EG - F^2$ can be derived from the above expressions. It can also be arranged as the product of two determinants:

$$EG - F^2 = \begin{vmatrix} E' & F' \\ F' & G' \end{vmatrix} \begin{vmatrix} \dfrac{\partial U}{\partial u} & \dfrac{\partial U}{\partial v} \\ \dfrac{\partial V}{\partial u} & \dfrac{\partial V}{\partial v} \end{vmatrix}^2 \quad . \tag{3.22}$$

The determinant

$$\begin{vmatrix} \dfrac{\partial U}{\partial u} & \dfrac{\partial U}{\partial v} \\ \dfrac{\partial V}{\partial u} & \dfrac{\partial V}{\partial v} \end{vmatrix}$$

is the Jacobean determinant of $(U, V)$ with respect to $(u, v)$.

## 3.3  Further concepts

### 3.3.1  *The scale distortion. Conditions of conformality and equivalency*

The scale distortion $m$ is defined by the ratio

$$m^2 = \frac{\mathrm{d}S^2}{\mathrm{d}s^2} = \frac{E' \, \mathrm{d}U^2 + 2F' \, \mathrm{d}U \, \mathrm{d}V + G' \, \mathrm{d}V^2}{e \, \mathrm{d}u^2 + 2f \, \mathrm{d}u \, \mathrm{d}v + g \, \mathrm{d}v^2}$$

or after some deductions

$$\left. \begin{aligned} m^2 &= \frac{E \, \mathrm{d}u^2 + 2F \, \mathrm{d}u \, \mathrm{d}v + G \, \mathrm{d}v^2}{e \, \mathrm{d}u^2 + 2f \, \mathrm{d}u \, \mathrm{d}v + g \, \mathrm{d}v^2} \\[2mm] &= \frac{E\left(\dfrac{\mathrm{d}u}{\mathrm{d}v}\right)^2 + 2F \dfrac{\mathrm{d}u}{\mathrm{d}v} + G}{e\left(\dfrac{\mathrm{d}u}{\mathrm{d}v}\right)^2 + 2f \dfrac{\mathrm{d}u}{\mathrm{d}v} + g} \quad . \end{aligned} \right\} \tag{3.23}$$

It is noted from this formula, that $m$ is dependent on the direction $\mathrm{d}u/\mathrm{d}v$ of the tangent at $P$ (Fig. 3.3) and is different in every direction. There is only one important exception namely if the coefficients of the denominator are respectively proportional to those in the numerator

$$\frac{E}{e} = \frac{F}{f} = \frac{G}{g} = m^2. \tag{3.24}$$

The scale distortion then is independent of $du/dv$ and is the same in every direction. This condition should be satisfied in all conformal projections. The same condition keeps the angles free of distortions in the image. This can be seen from formulae (3.15) and (3.16) in combination with (3.24) whence

$$\sin \omega = \sin \Omega$$

and

$$\sin \vartheta_1 = \sin \Theta_1 .$$

According to (3.18) the area of an infinitely small parallelogram was

$$A_D = \sqrt{eg - f^2} \; du \; dv.$$

If the image of this parallelogram should have the same area the following condition should be satisfied

$$eg - f^2 = EG - F^2 = \begin{vmatrix} E' & F' \\ F' & G' \end{vmatrix} \begin{vmatrix} \dfrac{\partial U}{\partial u} & \dfrac{\partial U}{\partial v} \\ \dfrac{\partial V}{\partial u} & \dfrac{\partial V}{\partial v} \end{vmatrix}^2 \tag{3.25}$$

taking a scale factor of one.

### 3.3.2 The fundamental Gaussian quantities on the ellipsoid, the sphere and in the plane; isometric coordinates

It has been seen in the previous chapter that the ellipsoid can be described by the equations

$$\left. \begin{aligned} x &= N \cos \varphi \cos \lambda \\ y &= N \cos \varphi \sin \lambda \\ z &= N(1 - \varepsilon^2) \sin \varphi. \end{aligned} \right\} \tag{3.26}$$

The parametric curves on the surface are the meridians and the *orthogonal* trajectories are the parallel circles ($u = \varphi$; $v = \lambda$).

The first fundamental quantities become:

$$\left.\begin{aligned} e &= M^2 \\ f &= 0 \\ g &= N^2 \cos^2 \varphi \end{aligned}\right\}, \tag{3.27}$$

and the equation for the elementary arc

$$ds^2 = M^2 \, d\varphi^2 + N^2 \cos^2 \varphi \, d\lambda^2. \tag{3.28}$$

The sphere is given by

$$\left.\begin{aligned} x &= R \cos \varphi \cos \lambda \\ y &= R \cos \varphi \sin \lambda \\ z &= R \sin \varphi, \end{aligned}\right\} \tag{3.29}$$

whence

$$\left.\begin{aligned} e &= R^2 \\ f &= 0 \\ g &= R^2 \cos^2 \varphi \end{aligned}\right\} \tag{3.30}$$

and

$$ds^2 = R^2 \, d\varphi^2 + R^2 \cos^2 \varphi \, d\lambda^2. \tag{3.31}$$

In the plane

$$dS^2 = dX^2 + dY^2. \tag{3.32}$$

It should be noted that

$$E' = G' = 1 \quad \text{and} \quad F' = 0, \tag{3.33}$$

the units of measure along the coordinate axes being equal. This is not so in the formulae (3.28) and (3.31). However they may be submitted to a transformation in order to make them equal. Condition is that $E/G$ is a quotient of a function of $u$ by a function of $v$

It was shown in (3.28) that

$$ds^2 = M^2 \, d\varphi^2 + N^2 \cos^2 \varphi \, d\lambda^2$$

or

$$ds^2 = N^2 \cos^2 \varphi \left\{ \left( \frac{M \, d\varphi}{N \cos \varphi} \right)^2 + d\lambda^2 \right\}. \tag{3.34}$$

A new variable is now introduced:

$$d\psi = \frac{M}{N \cos \varphi} d\varphi = \frac{(1+\varepsilon^2)}{(1-\varepsilon^2 \sin^2 \varphi) \cos \varphi} d\varphi \tag{3.35}$$

using

$$M = \frac{a(1-\varepsilon^2)}{W^3} \quad \text{and} \quad N = \frac{a}{W}.$$

This equation can be written as:

$$d\psi = \frac{(1-\varepsilon^2 \sin^2 \varphi - \varepsilon^2 \cos^2 \varphi)}{(1-\varepsilon^2 \sin^2 \varphi) \cos \varphi} d\varphi =$$

$$= \frac{1}{\cos \varphi} d\varphi - \frac{1}{2} \varepsilon \frac{2\varepsilon \cos \varphi}{(1-\varepsilon^2 \sin^2 \varphi)} d\varphi =$$

$$= \frac{1}{\cos \varphi} d\varphi + \frac{1}{2} \varepsilon \frac{-\varepsilon(1+\varepsilon \sin \varphi) \cos \varphi - \varepsilon(1-\varepsilon \sin \varphi) \cos \varphi}{\left(\dfrac{1-\varepsilon \sin \varphi}{1+\varepsilon \sin \varphi}\right)(1+\varepsilon \sin \varphi)^2}.$$

By integration it is found that

$$\psi = \ln \tan (45° + \tfrac{1}{2}\varphi) + \tfrac{1}{2}\varepsilon \ln \frac{1-\varepsilon \sin \varphi}{1+\varepsilon \sin \varphi} + \text{a constant.} \tag{3.36}$$

The constant however is equal to zero, taking $\psi = 0$ if $\varphi = 0$. Formula (3.34) now becomes

$$ds^2 = N^2 \cos^2 \varphi (d\psi^2 + d\lambda^2) \tag{3.37}$$

where $\bar{E} = \bar{G} = N^2 \cos^2 \varphi$ and $\bar{F} = 0$.

This new coordinate system $(\psi, \lambda)$ is called an *isometric* coordinate system* with isometric parameters [7], [9], [29].

The parameter $\psi$ is called the isometric latitude.

The isometric latitude on the sphere is derived directly by putting the excentricity $\varepsilon$ equal to zero in (3.36) giving

$$\psi_{\text{Sphere}} = \ln \tan (45° + \tfrac{1}{2}\varphi). \tag{3.38}$$

---

* Not every parameter system can be transformed into an isometric system. It depends on the possibility to change the parameters suitably, which in turn depends on the possibility of integration. Parameters that can be transformed in this way are called isothermal.

The line element then is

$$ds^2 = R^2 \cos^2 \varphi \, (d\psi^2 + d\lambda^2).$$ (3.39)

In a plane polar coordinate system, where the line element is given by

$$ds^2 = d\rho^2 + \rho^2 \, d\vartheta^2$$ (3.40)

($\rho$ is the polar ray and $\vartheta$ the polar angle) the fundamental quantities $e$ and $g$ are *not* equal to those in equation (3.32).

By transformation to an isometric system

$$ds^2 = \rho^2 \left( \frac{d\rho^2}{\rho^2} + d\vartheta^2 \right)$$

or

$$ds^2 = \rho^2 (dt^2 + d\vartheta^2)$$ (3.41)

where $dt = d\rho/\rho$. By integration $t = \ln \rho$.

In general if $(\bar{u}, \bar{v})$ and $(\overline{U}, \overline{V})$ are isometric systems

$$ds^2 = \frac{1}{p^2}(d\bar{u}^2 + d\bar{v}^2) \quad \text{and} \quad dS^2 = \frac{1}{P^2}(d\overline{U}^2 + d\overline{V}^2).$$ (3.42)

The quantity $p$ is called the *density* of the parametric curves. The distance between the curves changes inversely proportional to $p$.

In a rectangular coordinate system the elementary parallelograms are rectangles ($\omega = 90°$ and $f = 0$; see Figure 3.3). These become squares in an isometric system.

Isometric parameters are of significance in conformal projections. Although the fundamental quantities in the condition of conformality (4.24) can be derived from (3.21) directly, the introduction of isometric parameters sometimes simplifies the formulae considerably.

## 3.4   The theory of distortions of distances, angles and areas.

### 3.4.1   *Corresponding rectangular coordinate systems on datum and image planes*

If a rectangular system of parametric curves has been selected on the datum plane ($f = 0$), the corresponding set of curves on the image plane is as a rule non rectangular ($F \neq 0$). However Tissot [27] has proven that in every point of the datum plane a set of rectangular parametric curves exists which has a corresponding set of the same characteristics in the image plane. These are called the principal parametric curves.

The proof is relatively simple:

In a point $P$ of a curved surface there are an infinite number of orthogonal parametric curves and the tangents at $P$ intersect at right angles. Assume these tangents have a position as in Figure 3.4.

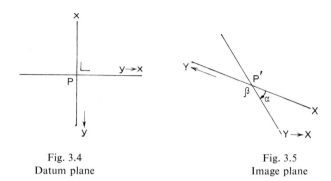

Fig. 3.4
Datum plane

Fig. 3.5
Image plane

In the image plane they will intersect as shown in Figure 3.5, where $X$ is the image of $x$ and $Y$ that of $y$, the angle $\alpha$ being the angle of intersection. After a rotation about $90°$ of $x$ and $y$, $x$ will have taken the position of $y$ and consequently $X$ the position of $Y$ in the image plane. The angle of intersection at $P'$ has changed from $\alpha$ into $\beta$ and has therefore gone through $90°$. Hence on the datum plane there is always one orthogonal set of parametric curves which is transformed into an orthogonal set on the image plane ($F = 0$).

These corresponding sets of principal curves will be referred to in all following derivations, if not stated otherwise (see Sections 3.5.3 and 5.1).

The formula for $m$ can now be simplified and becomes

$$m^2 = \frac{E \, du^2 + G \, dv^2}{e \, du^2 + g \, dv^2} \tag{3.43}$$

### 3.4.2    *The scale distortion on the parametric curves*

The scale distortion along the $u$ curve (where $dv = 0$) is equal to

$$m_0 = \sqrt{\frac{E}{e}}. \tag{3.44}$$

Similarly the scale distortion along the $v$ curve

$$m_{90} = \sqrt{\frac{G}{g}}. \tag{3.45}$$

Using (3.13) and (3.16) (omitting the index one for convenience),

$$\cos \vartheta = \sqrt{e}\,\frac{du}{ds} \quad \text{and} \quad \sin \vartheta = \sqrt{g}\,\frac{dv}{ds} \tag{3.46}$$

and (3.44) and (3.45)

$E = em_0^2$ and $G = gm_{90}^2$ respectively, equation (3.43) can be transposed into

$$m^2 = \frac{e\,du^2\,m_0^2 + g\,dv^2\,m_{90}^2}{ds^2}$$

or

$$m^2 = m_0^2 \cos^2 \vartheta + m_{90}^2 \sin^2 \vartheta. \tag{3.47}$$

The maximum and minimum scale distortion are obtained by putting $dm/d\vartheta = 0$

By differentiation of (3.47) it is found that

$$\frac{d(m^2)}{d\vartheta} = -2 \sin \vartheta \cos \vartheta\, m_0^2 + 2 \sin \vartheta \cos \vartheta\, m_{90}^2 = (m_{90}^2 - m_0^2) \sin 2\vartheta$$

Equating this expression to zero yields an extreme value $m$ for $\vartheta = 0$

$$m_{\text{extr}} = m_0 \tag{3.48}$$

and for $\vartheta = 90°$

$$m_{\text{extr}} = m_{90}. \tag{3.49}$$

This is an important feature because it appears that the directions of the maximum and minimum scale distortion are identical with the direction of the parametric $u$ and $v$ curves, and are therefore orthogonal.

### 3.4.3   The angular distortion; maxima and minima

The angular distortion on the image surfaces can be derived as follows:
The expressions $\cos \vartheta$ and $\sin \vartheta$ are given by (3.46).
On the projection surface

$$\cos \Theta = \sqrt{E}\,\frac{du}{dS} \quad \text{and} \quad \sin \Theta = \sqrt{G}\,\frac{dv}{dS} \tag{3.50}$$

(see Fig. 3.6).

Now $\sin (\Theta - \vartheta) = + \sin \Theta \cos \vartheta - \cos \Theta \sin \vartheta$

$$
\left.
\begin{aligned}
&= -\sqrt{g}\,\frac{dv}{ds}\sqrt{E}\,\frac{du}{dS} + \sqrt{e}\,\frac{du}{ds}\sqrt{G}\,\frac{dv}{dS} \\[4pt]
&= (\sqrt{eG} - \sqrt{gE})\,\frac{du}{ds}\frac{dv}{dS}
\end{aligned}
\right\}
\tag{3.51}
$$

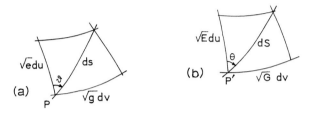

Fig. 3.6
(a) Datum surface, (b) projection surface

Combination of (3.44), (3.45) and (3.51) gives

$$
\sin (\Theta - \vartheta) = (m_{90} - m_0)\,\frac{du}{ds}\cdot\frac{dv}{dS}\sqrt{eg}\,.
\tag{3.52}
$$

Analogously it can be derived that

$$
\sin (\Theta + \vartheta) = (m_{90} + m_0)\,\frac{du}{ds}\cdot\frac{dv}{dS}\sqrt{eg}
\tag{3.53}
$$

whence

$$
\sin (\Theta - \vartheta) = \frac{m_{90} - m_0}{m_{90} + m_0}\sin (\Theta + \vartheta)\,.
\tag{3.54}
$$

By putting $\eta = \tfrac{1}{2}(\Theta + \vartheta)$, (3.54) becomes

$$
\sin (\Theta - \vartheta) = \frac{m_{90} - m_0}{m_{90} + m_0}\sin 2\eta\,.
\tag{3.55}
$$

The maximum distortion $\zeta = (\Theta - \vartheta)_{\max}$ of a bearing $\vartheta$ occurs when $\sin 2\eta = 1$ or when $\eta = 45°, 135°, 225°$ or $315°$.

Denoting the bearing of this distortion by $\bar{\vartheta}$ and $\bar{\Theta}$ respectively, it is seen that

and

$$
\left.
\begin{aligned}
\bar{\vartheta} &= 45^\circ - \tfrac{1}{2}\zeta = \eta_{(45^\circ)} - \tfrac{1}{2}\zeta \\[2mm]
\bar{\Theta} &= 45^\circ + \tfrac{1}{2}\zeta = \eta_{(45^\circ)} + \tfrac{1}{2}\zeta
\end{aligned}
\right\}
\tag{3.56}
$$

$$
\text{since} \quad \bar{\Theta} + \bar{\vartheta} = 90^\circ \quad \text{and} \quad \bar{\Theta} - \bar{\vartheta} = \zeta.
$$

If the angle $\eta$ increases $0 \leqslant \eta \leqslant 45^\circ$ the distortion increases accordingly until it reaches the maximum value $\zeta$, and decreases to zero for $45^\circ \leqslant \eta \leqslant 90^\circ$. If $90^\circ \leqslant \eta \leqslant 135^\circ$ the distortion increases again but with the opposite sign.

The maximum distortion is

$$
\sin \zeta = \sin (\Theta - \vartheta)_{\max} = \frac{m_{90} - m_0}{m_{90} + m_0} = (\Theta - \vartheta)_{\max}
\tag{3.57}
$$

$(\Theta - \vartheta)_{\max}$ being a small angle.

The distortion of an arbitrary angle – which can be considered as the difference of two bearings – can now be estimated.

In analogy to (3.55) the distortion of the second bearing $\vartheta_1$, of an angle $\delta = \vartheta - \vartheta_1$, is equal to

$$
\sin (\Theta_1 - \vartheta_1) = \frac{m_{90} - m_0}{m_{90} + m_0} \sin 2\eta_1
\tag{3.58}
$$

The angle $\delta$ has *no* distortion if

$$
\eta + \eta_1 = 90^\circ,
$$

both bearings $\vartheta$ and $\vartheta_1$ then having the same distortion in the same direction. An illustration is given in the Figure 3.7, where $\Theta - \vartheta = \Theta_1 - \vartheta_1$.

Condition is that $\eta$ and $\eta_1 < 90^\circ$.

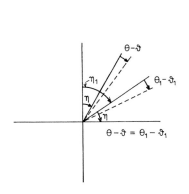

Fig. 3.7
Equal angular distortions

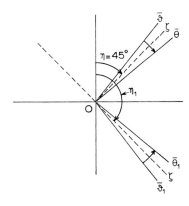

Fig. 3.8
Maximum angular distortions

The angle $\delta$ has a *maximum* distortion of $2\zeta$
if

$$\eta + \eta_1 = 180°$$

and in addition $\eta = 45°$. This is shown in Figure 3.8. Note that $\bar{9}0\bar{\Theta}_1$ should be equal to $90°$, *not* $\bar{9}0\bar{9}_1$

*Remark:* It can be seen from (3.52) again that $9 = \Theta$ if $m_0 = m_{90}$ as in all conformal projections.

### 3.4.4   *The scale distortion corresponding to the maximum angular distortion*

One may require the scale distortion $m_{\bar{9}}$ for the bearing $\bar{9}$ of the maximum directional distortion $\zeta$.
According to (3.56)

$$\bar{9} = 45° - \tfrac{1}{2}\zeta$$

whence in combination with (3.47)

$$m_{\bar{9}}^2 = m_0^2 \cos^2 (45° - \tfrac{1}{2}\zeta) + m_{90}^2 \sin^2 (45° - \tfrac{1}{2}\zeta) \tag{3.59}$$

Since

$$\sin \zeta = \frac{m_{90} - m_0}{m_{90} + m_0} \tag{see 3.57},$$

$$\cos (90° - \zeta) = \cos^2 (45° - \tfrac{1}{2}\zeta) - \sin^2 (45 - \tfrac{1}{2}\zeta) = -\frac{m_{90} - m_0}{m_{90} + m_0}. \tag{3.60}$$

Also

$$\cos^2 (45° - \zeta) + \sin^2 (45° - \tfrac{1}{2}\zeta) = 1. \tag{3.61}$$

Elimination of $\cos (45° - 2\zeta)$ and $\sin (45° - 2\zeta)$ from the equations (3.59), (3.60) and (3.61) yields

$$m_{\bar{9}}^2 = m_0 m_{90} \tag{3.62}$$

making the scale distortion $m_{\bar{9}}$ for the bearing of maximum directional distortion $\bar{9}$ equal to $\sqrt{m_0 m_{90}}$.

By recollecting (3.60) it follows that in equal area projections distances in these directions are depicted at their true length ($m_{\bar{9}} = 1$). See Section 6.1.2.

### 3.4.5   *The distortion of areas*

*The distortion of areas* is derived from (3.18) combined with (3.44) and (3.45):

$$\frac{A_P}{A_D} = \sqrt{\frac{EG}{eg}} = m_0 m_{90} \tag{3.63}$$

For equivalent or equal area projections $A_P/A_D = 1$ whence the important relationship

$$m_0 \, m_{90} = 1 \qquad (3.64)$$

## 3.5  Tissot's indicatrix. Derivation and application

### 3.5.1  Theory of the indicatrix

Tissot's theory of distortions states that a circle on the datum surface with a centre $P$ and a radius d$s$ may be assumed to be a plane figure within its infinitely small area. This area will remain infinitely small and plane on the projection surface. Hence the rules of the projective Euclidean geometry apply. Generally the circle will be portrayed as an ellipse. Only in the particular case that both datum and image planes are parallel the circle will remain a circle, though at a different scale. The elements of the ellipse are related to those of the circle by the formulae of an affine transformation.

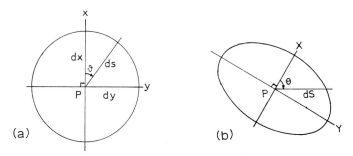

Fig. 3.9
(a) Datum plane, (b) projection plane

Consider the circle at $P$ on a rectangular coordinate system $x, y$ and the projection on the corresponding rectangular $X, Y$ system. The coordinate axes are therefore the tangents to the corresponding $(u, v)$ systems on both surfaces (Fig. 3.9). From (3.46) follows:

$$\left. \begin{array}{l} dx = ds \, \cos \vartheta = \sqrt{e} \; du \\[2mm] dy = ds \, \sin \vartheta = \sqrt{g} \; dv \end{array} \right\} \qquad (3.65)$$

and correspondingly

$$\left. \begin{array}{l} dX = dS \, \cos \Theta = \sqrt{E} \; du \\[2mm] dY = dS \, \sin \Theta = \sqrt{G} \; dv. \end{array} \right\} \qquad (3.66)$$

By substitution of

$$du = \frac{1}{\sqrt{e}} \cos \vartheta \, ds$$

and

$$dv = \frac{1}{\sqrt{g}} \sin \vartheta \, ds$$

(according to (3.13) and (3.16)), (3.66) becomes

$$dX = \sqrt{\frac{E}{e}} \cos \vartheta \, ds$$

and

$$dY = \sqrt{\frac{G}{g}} \sin \vartheta \, ds$$

or, using (3.44) and (3.45)

$$\left. \begin{array}{l} dX = m_0 \, \cos \vartheta \, ds \\ dY = m_{90} \sin \vartheta \, ds. \end{array} \right\} \tag{3.67}$$

By squaring $dX/m_0$ and $dY/m_{90}$ and their addition it is seen that

$$ds^2 = \frac{dX^2}{m_0^2} + \frac{dY^2}{m_{90}^2}. \tag{3.68}$$

This is the equation of an ellipse with a semi major (minor) axis $m_0 \, ds$ and a semi minor (major) axis $m_{90} \, ds$.

If $ds$ is taken as a unit radius the equation becomes

$$\frac{dX^2}{m_0^2} + \frac{dY^2}{m_{90}^2} = 1. \tag{3.69}$$

This ellipse is called *Tissot's indicatrix*, because it indicates the characteristics of a projection in the direct environment of a point $P$.

It has been shown in the previous chapter that the area on the earth that can be considered plane has a radius of approximately 10 km, and may even be larger according to the requirement of precision becoming less stringent. Within this region one may write equation (3.69) on a $X, Y$ regular coordinate system omitting the differentials:

$$\frac{X^2}{m_0^2} + \frac{Y^2}{m_{90}^2} = 1. \tag{3.70}$$

### 3.5.2  *Application of the indicatrix*

In using the indicatrix one may bring both coordinate systems $xy$ and $XY$ into coincidence, the circle and ellipse being concentric. This is illustrated in Figure 3.10.

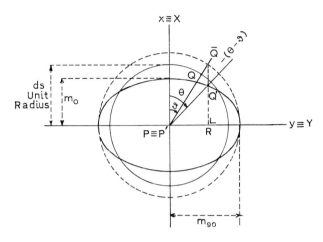

Fig. 3.10
Tissot's indicatrix

In this figure $Q'$ is the image of $Q$. Further

$$P\bar{Q} = m_{90}\ PQ = m_{90}\ ds.$$

Now

$$PR = Y_{\bar{Q}} = Y_{Q'} = m_{90}\ ds\sin\vartheta,$$

but since $ds$ is the unit radius

$$PR = Y_{Q'} = m_{90}\sin\vartheta. \tag{3.71}$$

Substitution into the equation (3.70) of the ellipse gives

$$X_{Q'} = m_0\cos\vartheta. \tag{3.72}$$

Also

$$RQ' = R\bar{Q}\frac{m_{90}}{m_0}.$$

Hence the coordinates of $Q'$ become (since $\sin\vartheta = X_Q/ds$ and $\cos\vartheta = Y_Q/ds$)

$$X_{Q'} = m_0\ X_Q$$

$$Y_{Q'} = m_{90}\ Y_Q$$

whence

$$\tan \Theta = \frac{m_{90}}{m_0} \tan \vartheta. \tag{3.73}$$

The other formulae (3.47), (3.57), (3.60) can be derived from Figure 3.10 by plane trigonometry.

### 3.5.3   Relationship of the elements of the indicatrix and the parametric latitude and longitude

In practice the parametric $(u, v)$ coordinate system does not always coincide with meridians and parallels of latitude $(\varphi, \lambda)$ on the ellipsoid and the sphere. In the case it does not, it is of interest to know the relationship of the elements of the indicatrix, if the $(u, v)$ system is rotated about an angle $\vartheta$ with respect to the system formed by the major and minor axes of the indicatrix. On the datum plane both systems are assumed to be rectangular. The $(u, v)$ system, however, will not remain rectangular but will intersect at an angle $\Omega$. The $m_\lambda$ and $m_\varphi$ in the indicatrix will intersect at the same angle. In the figure 3.11 the $(\varphi, \lambda)$ or $(u, v)$ system transforms into the $(\Phi, \Lambda)$ system, $P'A' = m_\varphi$; $P'B' = m_\lambda$. The angle $\vartheta$ transforms into $\Theta$. Further $\measuredangle APB = 90°$ and $\measuredangle A'P'B' = \Omega$.

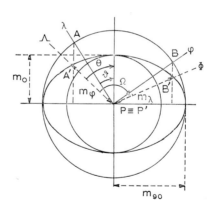

Fig. 3.11
Parametric latitude and longitude in indicatrix

It can be read from this figure directly that

$$m_0 \cos \vartheta = m_\varphi \cos \Theta \tag{3.74}$$

$$m_{90} \sin \vartheta = m_\varphi \sin \Theta \tag{3.75}$$

$$m_0 \sin \vartheta = m_\lambda \cos (\Omega - \Theta) \tag{3.76}$$

$$m_{90} \cos \vartheta = m_\lambda \sin (\Omega - \Theta). \tag{3.77}$$

The $m_0$, $m_{90}$ and $\vartheta$ are the unknowns in these equations.

By squaring and adding one obtains the important relationship

$$m_{90}^2 + m_0^2 = m_\lambda^2 + m_\varphi^2.$$  (3.78)

By multiplication of (3.74) and (3.77)

$$m_0\, m_{90} \cos^2 \vartheta = m_\lambda\, m_\varphi \cos \Theta \sin (\Omega - \Theta).$$  (3.79)

Similarly from (3.75) and (3.76)

$$- m_{90}\, m_0 \sin^2 \vartheta = m_\lambda\, m_\varphi \sin \Theta \cos (\Omega - \Theta).$$  (3.80)

Subtraction of this expression from (3.79) shows ultimately

$$m_{90}\, m_0 = m_\lambda\, m_\varphi \sin \Omega.$$  (3.81)

Application of (3.78) and (3.81) yields

$$\left.\begin{aligned}
(m_{90} + m_0)^2 &= m_\lambda^2 + m_\varphi^2 + 2\, m_\lambda\, m_\varphi \sin \Omega \\
(m_{90} - m_0)^2 &= m_\lambda^2 + m_\varphi^2 - 2\, m_\lambda\, m_\varphi \sin \Omega.
\end{aligned}\right\}$$  (3.82)

Also, since according to (3.47)

$$m_\lambda^2 = m_0^2 \cos^2 \vartheta_\lambda + m_{90}^2 \sin^2 \vartheta_\lambda\,,$$

$$\left.\begin{aligned}
\sin^2 \vartheta_\lambda &= \frac{m_0^2 - m_\lambda^2}{m_0^2 - m_{90}^2} \\[2ex]
\cos^2 \vartheta_\lambda &= \frac{m_\lambda^2 - m_{90}^2}{m_0^2 - m_{90}^2}.
\end{aligned}\right\}$$  (3.83)

and

The orientation angle of the axes of the indicatrix on the ellipsoid may also be calculated as follows.

Taking the scale distortion as

$$m^2 = \frac{E d\varphi^2 + 2\sqrt{EG} \cos \Omega\, d\lambda + G d\lambda^2}{e d\varphi^2 + g d\lambda^2}$$  (3.84)

the angle $\vartheta$ is required for which this expression is maximum (or minimum). Since $d\lambda / d\varphi = \sqrt{e/g}\, \tan \vartheta$, equation (3.84) may be written as

$$m^2 = \frac{m_\varphi^2 + 2 m_\varphi\, m_\lambda \cos \Omega \tan \vartheta + m_\lambda^2 \tan^2 \vartheta}{1 + \tan^2 \vartheta}.$$  (3.85)

Putting $\partial m^2/\partial\vartheta = 0$, one obtains the second degree equation

$$\tan^2\vartheta - \frac{m_\lambda^2 - m_\varphi^2}{m_\varphi m_\lambda \cos\Omega}\tan\vartheta - 1 = 0 \tag{3.86}$$

whence

$$\tan\vartheta = \frac{m_\lambda^2 - m_\varphi^2}{2 m_\varphi m_\lambda \cos\Omega} \pm \frac{1}{2}\sqrt{\frac{(m_\lambda^2 - m_\varphi^2)^2}{m_\varphi^2 m_\lambda^2 \cos^2\Omega} + 4}. \tag{3.87}$$

The corresponding angle $\Theta$ in the projection is equal to

$$\tan\Theta = \frac{\tan\vartheta}{m_\varphi^2 + \tan\vartheta\cot\Omega} \tag{3.88}$$

the proof being given in Section 6.1.2.

A formal example is also given in that section. For the general application of the Tissot indicatrix utilizing the electronic computer refer to Chapter 7.

### 3.5.4   *The indicatrix in practice. Simple cases*

It has been shown that a circle of unit radius $ds = 1$ on the part of the datum surface that can be replaced by the local tangential plane, is in general projected as an ellipse, called the indicatrix. The semi major and minor axes are equal to the maximum and minimum scale distortion respectively. These are calculated by (3.44) and (3.45) respectively

$$m_0 = \sqrt{\frac{E}{e}}$$

$$m_{90} = \sqrt{\frac{G}{g}}.$$

The maximum distortion $\zeta$ of a bearing is then given by (3.57)

$$\sin\zeta = \frac{m_{90} - m_0}{m_{90} + m_0}$$

and the maximum distortion of an angle is equal to $2\zeta$.

The scale distortion in an arbitrary bearing $\vartheta$ is according to (3.47)

$$m_\vartheta^2 = m_0^2\cos^2\vartheta + m_{90}^2\sin^2\vartheta$$

with specifically the scale distortion in the bearing $\bar\vartheta$ of the maximum distortion (see 3.62).

$$m_{\bar\vartheta} = \sqrt{m_0 m_{90}}.$$

The distortion of an area follows from (3.63)

$$\frac{A_P}{A_D} = m_0\, m_{90}.$$

The characteristic quantities are simple to calculate in the case of the less complicated geometrical projections of the sphere.

At a point $P(\varphi, \lambda)$ on the sphere the radius of the parallel circle at $P$ is equal to $R \cos \varphi$.

An increase $d\varphi$ of the latitude corresponds to a linear displacement along the meridian of $R\, d\varphi$. The radius of the corresponding parallel circle will change by $d(R \cos \varphi) = -R \sin \varphi$. In the projection the radius of this parallel circle (indicated by $\rho$, a function of the latitude $\varphi$) will change by $d\rho = f'(\varphi)$. The scale distortion $m_0$ therefore becomes

$$m_0 = \frac{d\rho}{R\, d\varphi} \qquad\qquad (3.89)$$

In a similar way it can be reasoned that

$$m_{90} = \frac{\rho}{R \cos \varphi}. \qquad\qquad (3.90)$$

The images of the meridian and parallel at $P'$ also intersect at right angles. Hence the $m_0$ and $m_{90}$ have the same directions as this meridian and parallel. Refer also to Chapter 7.

EXAMPLE

The stereographic projection treated in Chapter 4 and 5. The parallel circle of $P(\varphi, \lambda)$ is projected on the tangential plane through $NP$, in the way indicated in Figure 3.12. It becomes a circle of radius $\rho = 2R \tan (45° - \tfrac{1}{2}\varphi)$ as is easily derived.

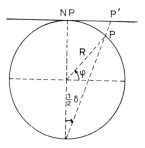

Fig. 3.12
Polar stereographic projection

The circumference in the datum plane is $2\pi R \cos \varphi$; in the projection it becomes equal to $4\pi R \tan (45° - \frac{1}{2}\varphi)$.

Now using (3.89) and (3.90)

$$m_0 = \frac{d\rho}{R\,d\varphi} = \frac{1}{\cos^2 (45° - {}_{\angle}\,\varphi)}$$

$$m_{90} = \frac{2R \tan (45° - \frac{1}{2}\varphi)}{2R \sin (45° - \frac{1}{2}\varphi) \cos (45° - \frac{1}{2}\varphi)} = \frac{1}{\cos^2 (45° - \frac{1}{2}\varphi)}.$$

Whence $m_0 = m_{90}$ (conformality!).

Further

$$\zeta = 0,$$

$$\frac{A_{P'}}{A_P} = \frac{1}{\cos^4 (45° - \frac{1}{2}\varphi)}.$$

Thus for

$$\varphi = 90°\quad m = 1$$

$$\varphi = 85°\quad m_0 = m_{90} = 1.002$$

$$\varphi = 75°\quad m_0 = m_{90} = 1.017\quad \text{etc.}$$

See Figure 3.13.

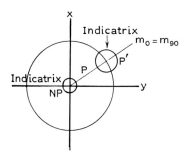

Fig. 3.13
Indicatrix of the projection

CHAPTER 4

AZIMUTHAL AND EQUIDISTANT PROJECTIONS

## 4.1 Azimuthal Projections – Introduction

Azimuthal projections have certain characteristics specific to this class, namely that they are theoretically or actually "projected" on a plane from the spherical datum surface. They are often called perspective projections, the term originating from the generation process employed. All these projections may be visualized as projected upon a plane perpendicular to a line passing through the center of the datum sphere or a plane perpendicular to an ellipsoidal normal. Thus, they may also be visualized as a photograph of the datum surface taken by an aircraft, a rocket or a space vehicle of the earth, the moon or another heavenly body, as suggested by Laubscher [12], the camera axis being coincident with the direction of the normal to the datum surface. If the projection plane is made tangent to the datum surface, there is no deformation of any kind at the center and furthermore, in such a case all great circles passing through the point of tangency will be straight lines on the projection surface, showing correct azimuths from the center to any point, hence the name azimuthal projections.

Perspective projections are based on simple geometric principles. The projection axis is perpendicular to both the datum and the projection surfaces and the projection center lies at the intersection of this axis and the projection surface. A point in space on the same axis serves as the perspective point and straight lines from the perspective point through the datum surface and the projection surface locate the respective surface points.

A parallel displacement of the projection plane along the axis changes only the scale of the projection, while the location of the perspective point changes the form of the projection, thus becoming the factor which determines the characteristics of a particular azimuthal (perspective) projection. There are three varieties of the azimuthal projections depending on the attitude of the projection axis. If the projection axis coincides with the rotation axis of the datum body, we have a polar or normal projection. If the projection axis is in the equatorial plane of the datum, we have a transverse projection. All other attitudes of the axis result in an oblique projection.

### 4.1.1  *The gnomonic projection*

The gnomonic projection is perhaps the simplest of the azimuthal projections. Using a sphere as the datum surface, the perspective point is placed at the center of the sphere and the projection is achieved through rays from this point, intersecting the spherical surface and projecting it onto a plane tangent to the sphere.

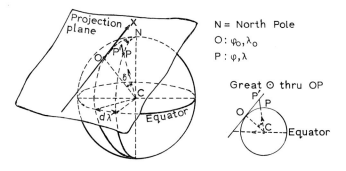

N = North Pole

$O: \varphi_0, \lambda_0$

$P: \varphi, \lambda$

Great $\odot$ thru OP

Fig. 4.1
The general case of gnomonic projection

The projection plane is made tangent to the sphere at $O$ (latitude $\varphi_0$), which becomes the center of the projection. The perspective point $C$ is located at the center of the datum sphere of radius $R$.

$P$ is an arbitrary point on the datum surface at latitude $\varphi$ and at longitude difference of $d\lambda$ from $O$, projected onto the projection surface at $P'$. It should be noted that $CO = CP = R$. The angle $\delta$ is subtended at $C$ between the radii $CO$ and $CP$. There is a system of rectangular coordinates in the projection plane as illustrated in Figure 4.2, showing the position $P'$ of the projected point $P$ within the plane rectangular system.

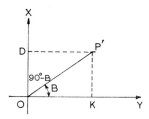

Fig. 4.2
Plane rectangular system of the projection

Noting that the North projection of the meridian through $O$ coincides with the $X$ axis and the $Y$ axis is the projection of great circle through $O$ which is perpendicular to the meridian through $O$, we have the following relationships:

Angle $NOP = 90 - B$

Arc    $NO \;= 90 - \varphi_0$

Arc    $NP \;= 90 - \varphi$ .

From Figure 4.1, it transpires that

$$OP' = R \tan \delta \qquad (4.1)$$

Utilizing figures 4.1 and 4.2 the following is derived:

$$X = EP' - OP' = OP' \sin B = R \tan \delta \sin B = \frac{R \sin \delta \sin B}{\cos \delta} \qquad (4.2)$$

$$Y = DP' = OP' \cos B = R \tan \delta \cos B = \frac{R \sin \delta \cos B}{\cos \delta}. \qquad (4.3)$$

From the spherical triangle $OPN$ in Figure 4.2 by sine law

$$\frac{\sin d\lambda}{\sin \delta} = \frac{\sin (90 - B)}{\sin (90 - \varphi)} = \frac{\cos B}{\cos \varphi}$$

and thus

$$\sin \delta \cos B = \sin d\lambda \cos \varphi \qquad (4.4)$$

and from the same triangle by cosine law

$$\sin \delta \sin B = \cos \varphi_0 \sin \varphi - \sin \varphi_0 \cos \varphi \cos d\lambda \qquad (4.5)$$

$$\cos \delta = \sin \varphi_0 \sin \varphi + \cos \varphi_0 \cos \varphi \cos d\lambda . \qquad (4.6)$$

Now substituting (4.4), (4.5) and (4.6) into (4.2) and (4.3) we obtain the mapping equations for the gnomonic projection.

$$X = \frac{R (\cos \varphi_0 \sin \varphi - \sin \varphi_0 \cos \varphi \cos d\lambda)}{\sin \varphi_0 \sin \varphi + \cos \varphi_0 \cos \varphi \cos d\lambda} \qquad (4.7)$$

$$Y = \frac{R \cos \varphi \sin d\lambda}{\sin \varphi_0 \sin \varphi + \cos \varphi_0 \cos \varphi \cos d\lambda}. \qquad (4.8)$$

The mapping equations (4.7) and (4.8) permit the computation of plane rectangular coordinates of the projection for any point on the datum surface given the point's latitude and longitude $d\lambda$ relative to the meridian through the projection center $O$.

The other quantities required are the radius of the datum sphere $R$ and the latitude of the projection center $\varphi_0$.

The mapping equations are easily programmable for the electronic computer and by reading in the latitudes and longitudes of the points to be mapped consecutively, the output will provide plotting data.

It is particularly advantageous to arrange the output in a form suitable for an automatic plotter, which performs such operations speedily and efficiently.

The simplest form of the gnomonic projection presented above as a general case is the gnomonic polar projection in which the projection plane is tangent at the pole (see Fig. 4.3).

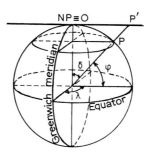

Fig. 4.3
The polar gnomonic projection

Noting that the projection center $O'$ is now at the North Pole, the latitude $\varphi_0 = 90°$ and referring the longitudes to the Greenwich meridian, the general mapping equations (4.7) and (4.8) can be modified as follows:

$$X = -\frac{R \cos \lambda \cos \varphi}{\sin \varphi} = -R \cos \lambda \cot \varphi = -R \tan \delta \cos \lambda \qquad (4.9)$$

$$Y = \frac{R \cos \varphi \sin \lambda}{\sin \varphi} = R \sin \lambda \cot \varphi = R \tan \delta \sin \lambda. \qquad (4.10)$$

The minus sign in formula (4.9) indicates that the orientation of the plane rectangular coordinate system is as shown in Figure 4.4.

The length distortions along the meridians are represented by the semi major axis of the Tissot indicatrix (see formulae (3.89) and (3.90)).

$$m_0 = \frac{R \tan \delta \, d\lambda}{R \sin \delta \, d\lambda} = \frac{1}{\cos \delta} = \frac{1}{\sin \varphi}. \qquad (4.11)$$

Also

$$m_{90} = \frac{dS}{ds} = \frac{1}{\sin^2 \varphi}. \qquad (4.12)$$

Since the indicatrix axes are not equal, (except in the origin $O$) it is immediately obvious that the projection is not conformal and since their product does not equal unity, the projection is not equivalent (equal areas). It can be described according to the classification in Chapter 1 as: Azimuthal, normal, tangential, geometric and regular (non-conventional).

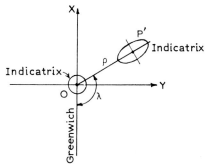

Fig. 4.4
The plane rectangular system of the polar gnomonic projection

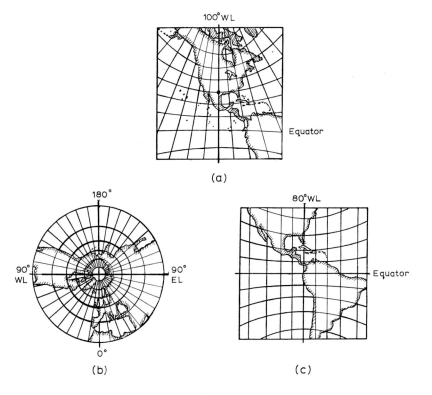

(a)

(b)                      (c)

Fig. 4.5
Gnomonic charts. (a) Oblique gnomonic projection, (b) normal gnomonic projection, (c) transverse gnomonic projection

The gnomonic projection has long been known, but it is now enjoying a return to popularity for representations accentuating the point coinciding with the projection's center, where there is no distortion of directions. This may be particularly attractive for, say, a national airline basing its services on a major airport which would become the projection's center and showing its radiating principal flight routes. It may also be attractive for mapping the polar regions. A complete hemisphere cannot be shown on a map plotted on the gnomonic projection since the hemisphere boundary would be at infinity from the center (equator would be at infinity in the polar gnomonic), but a straight line drawn between any two points would represent an arc of a great circle between them and is extremely attractive for showing the localities through which the most direct route passes. This finds its application in great circle flying or sailing and charts based on the gnomonic projection have been published covering in single sheets such routes as the North Atlantic, The Polar, the South Pacific and others. For the graphical construction principles and procedures the reader is referred to [23].

### 4.1.2   *The stereographic projection*

The stereographic projection is very similarly developed to the gnomonic, the major difference being that in the stereographic projection the perspective point is placed diametrically opposed to the point of tangency of the projection plane with the datum surface.

Figure 4.6 illustrates this generation feature.

The derivation of mapping equations is made along the same lines as those for the gnomonic.

Figure 4.6 shows the basic feature that the perspective point is diametrically

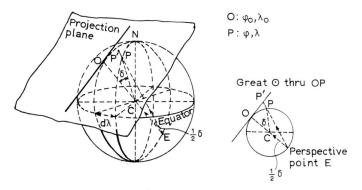

Fig. 4.6
The general case of stereographic projection

opposite the point of tangency and thus $CE = OC = R$. The angle $ONP$ is again the difference of longitude $d\lambda$ between the meridian through $O$ and the meridian through a random point $P$ projected onto the projection plane at $P'$. The plane rectangular coordinate system of the projection is oriented in the same manner as in the gnomonic (see Fig. 4.2).

With the aid of Figures 4.2 and 4.6, the following may be derived:

$$X' = KP' = OP' \sin B = 2R \tan \frac{\delta}{2} \sin B = \frac{2R \sin \delta \sin B}{1 + \cos \delta} \qquad (4.13)$$

$$Y' = DP' = OP' \cos B = 2R \tan \frac{\delta}{2} \cos B = \frac{2R \sin \delta \cos B}{1 + \cos \delta}. \qquad (4.14)$$

Utilizing the spherical triangle $OPN$ the equations (4.4), (4.5) and (4.6) are again derived and substituted into (4.13) and (4.14) resulting in the following mapping equations for the stereographic projection:

$$X = \frac{2R (\cos \varphi_0 \sin \varphi - \sin \varphi_0 \cos \varphi \cos d\lambda)}{1 + \sin \varphi_0 \sin \varphi + \cos \varphi_0 \cos \varphi \cos d\lambda} \qquad (4.15)$$

$$Y = \frac{2R \cos \varphi \sin d\lambda}{1 + \sin \varphi_0 \sin \varphi + \cos \varphi_0 \cos \varphi \cos d\lambda}. \qquad (4.16)$$

The simplest form of the stereographic projection is the normal case known as the polar stereographic. In this projection the projection plane is tangent at one of the poles, with the perspective point at the other pole (see Fig. 4.7).

In the above illustrated case, $\varphi_0 = 90°$ and $\delta = 90° - \varphi$.

Modifying the general case's mapping equations (4.15) and (4.16) we obtain:

$$X = -\frac{2R \cos \varphi \cos d\lambda}{1 + \sin \varphi} = \frac{2R \sin \delta \cos d\lambda}{1 + \cos \delta} = 2R \tan \frac{\delta}{2} \cos d\lambda \quad (4.17)$$

$$Y = \frac{2R \cos \varphi \sin d\lambda}{1 + \sin \varphi} = \frac{2R \sin \delta \sin d\lambda}{1 + \cos \delta} = 2R \tan \frac{\delta}{2} \sin d\lambda. \quad (4.18)$$

The minus sign in formula (4.17) indicates that the orientation of the plane rectangular coordinate system is as shown in Fig. 4.8. If the longitude is referred to the Greenwich meridian, as in Fig. 4.7 and 4.8, the $d\lambda$ in (4.17) and (4.18) may be replaced by $\lambda$.

The length distortions along the meridians are represented by the semi axis of the Tissot indicatrix.

$$m_0 = \frac{1}{\cos^2 \frac{\delta}{2}} = \frac{1}{\cos^2 (45° - \frac{1}{2}\varphi)}. \qquad (4.19)$$

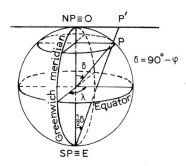

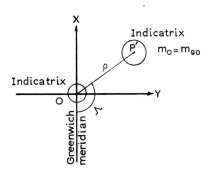

Fig. 4.7
The polar stereographic projection
for the north polar area

Fig. 4.8
The plane rectangular system
of the polar stereographic projection

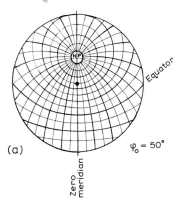

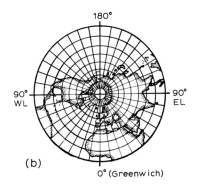

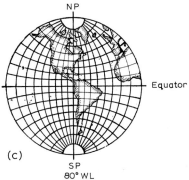

Fig. 4.9
(a) Oblique stereographic projection, (b) normal stereographic projection ($NP = O$), (c) transverse stereographic projection. The stereographic projection has the special property, that a circle on the datum surface remains circular in the projection

The length distortions along the parallels are represented by the other semi axis of the indicatrix.

$$m_{90} = \frac{1}{\cos^2 \dfrac{\delta}{2}} = \frac{1}{\cos^2 (45° - \frac{1}{2}\varphi)}. \tag{4.20}$$

The identical expressions (4.19) and (4.20) show that $m_0 = m_{90}$ and thus the projection is conformal. The meridians are straight lines and there is no angular distortion at any point (differentially), the Tissot indicatrix being a circle.

The three types of stereographic grids are shown in Figure 4.9.

### 4.1.3 The orthographic projection

The orthographic projection is obtained by placing the perspective point at infinity and thus forming parallel projection rays which project the datum surface on a plane tangent to the datum and perpendicular to the projecting rays (see Fig. 4.10).

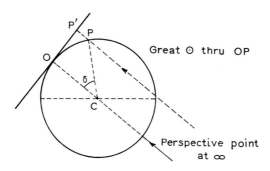

Fig. 4.10
Orthographic projection

The distance $PO'$, i.e. the distance from the center of projection and the projected point is

$$OP' = R \sin \delta . \tag{4.21}$$

Utilizing the same rectangular system as in Fig. 4.2.

$$X' = EP' = OP' \sin B = R \sin \delta \sin B \tag{4.22}$$

$$Y' = DP' = OP' \cos B = R \sin \delta \cos B. \tag{4.23}$$

Equations (4.4) and (4.5) are again derived and substituting those into (4.22) and (4.23) the final mapping equations for the orthographic projection are

$$X = R \left( \cos \varphi_0 \sin \varphi - \sin \varphi_0 \cos \varphi \cos d\lambda \right). \tag{4.24}$$

$$Y = R \cos \varphi \sin d\lambda. \tag{4.25}$$

With the perspective point at infinity and the radii of the parallels on the projection proportional to the radii of the parallel circles on the datum, the scale along the parallels is held true. The effect of the orthographic projection is like a photograph of the datum surface grid taken from a great distance. This property is useful for representations stressing the visual aspect such as viewing the moon from the earth or say, viewing the Antarctic from the North etc. (see Fig. 4.11).

The application of formulae (3.89) and (3.90) shows that the semi axes of the

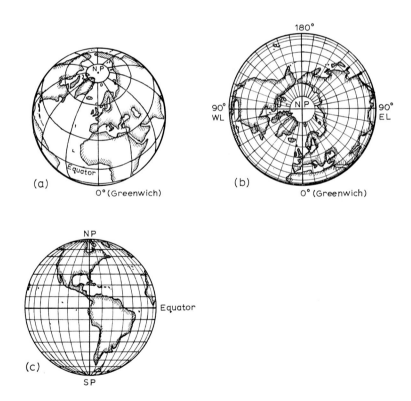

Fig. 4.11
Orthographic projections. (a) Oblique orthographic projection, (b) normal orthographic projection, (c) transverse orthographic projection

indicatrix in the oblique orthographic projection are

$$\left.\begin{array}{l} m_0 = \cos\delta \\ m_{90} = 1 \end{array}\right\} \qquad (4.26)$$

with

$$\sin\zeta = \tan^2 \tfrac{1}{2}\delta .$$

In the normal case $\delta = 90 - \varphi$. In the origin $m_0 = m_{90} = 1$. The rectangular system and the shape of the indicatrix is shown in Figure 4.12.

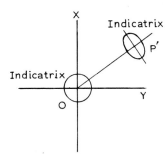

Fig. 4.12
Indicatrix of the projection

### 4.1.4    The azimuthal equidistant projection

An azimuthal equidistant projection, frequently used for setting out bearing and distance with respect to a selected origin is "Postel's projection".

The great circles through the origin are depicted as straight lines, whilst the true arc length is set off along those lines. This is illustrated in Fig. 4.13 showing the plane of the great circle through $O$ and $P$ where arc $OP = $ arc $\delta = OP'$.

The axes of the indicatrix are obtained in the usual manner

$$\left.\begin{array}{l} m_0 \;\; = 1 \\ m_{90} = \dfrac{\text{arc }\delta}{\sin\delta} . \end{array}\right\} \qquad (4.27)$$

Since $m_0 m_{90} \neq 1$ and $m_0 \neq m_{90}$ (except in the origin); the projection is neither equivalent nor conformal.

The plane coordinate system and the indicatrix are shown in Figure 4.14, where $X$ is directed towards the North projection of the meridian at $O$. $OP' = \rho = $ arc $\delta$. Two examples are shown in Figure 4.15.

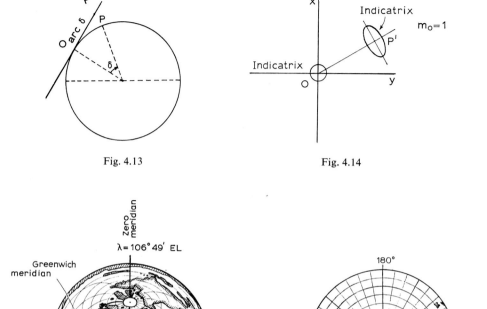

Fig. 4.13                              Fig. 4.14

Fig. 4.15
(a) After [19]. Oblique azimuthal equidistant projection, (b) polar azimuthal equidistant
projection

The plane angles at $O$ are equal to the spherical angles at that point.

### 4.1.5   *The azimuthal equivalent projection*

The Lambert azimuthal equivalent projection is treated in Chapter 6.

The length $\rho$ set off along the projection of the great circles radiating from the central point is equal to the chord

$$\rho = 2R \sin \tfrac{1}{2} \delta .$$ (4.28)

The axes of the indicatrix then are

$$m_0 = \cos \tfrac{1}{2}\delta$$

$$\left.\begin{array}{l} m_0 = \cos \tfrac{1}{2}\delta \\[2ex] m_{90} = \dfrac{1}{\cos \tfrac{1}{2}\delta} \end{array}\right\}$$  (4.29)

with $m_0 m_{90} = 1$.

### 4.1.6   *Comparison of grids*

A comparison of the various grids of the projections discussed in this chapter is shown in Figure 4.16.

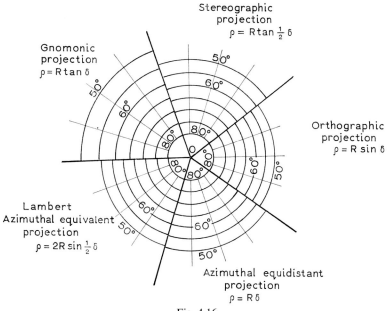

Fig. 4.16
Comparison of grids

## 4.2.   The conical and cylindrical equidistant projections ;
## The Cassini-Soldner projection

The equidistant projections utilize the cone and the cylinder in the normal and transverse attitudes, achieving the equidistancy along the line of contact. Thus in the normal cases there is a standard parallel where the distances are preserved between the intersections with the meridians. Correspondingly in the transverse cases there is a standard meridian along which equidistancy exists between its intersections with the parallels.

There is a fairly rapid deterioration of scale away from the standard circles and thus the normal cases are applicable to narrow regions of considerable

East-West extent and the transverse cases to narrow regions extending in the North-South direction. Whilst the scale deterioration in the cylindrical equidistant projections is symmetrical about the standard circle, the conical equidistant projections show no such symmetry and are consequently of very little use except in some atlas presentations.

The cylindrical equidistant projections are in use by several countries where the narrow shape is divided by either the standard meridian or the standard parallel as the case may be. The Cassini–Soldner projection is employed in some countries of predominant North–South extent such as Great Britain. This projection is classified as cylindrical, tangent, transverse, equidistant and semi-geometric (see 1.3.3). The cylinder is tangent along the meridian centrally situated, so that the narrow belt is bisected by it as symmetrically as possible. The North–South grid lines of the projection system are projections of small circles, parallel to the central meridian. The distances along the $X$ axis in the projection are always greater than the corresponding arc distances on the datum surface. If a spherical datum surface is used, equidistancy is preserved along the central meridian, but this is not so for an ellipsoidal datum.

The $Y$ coordinates of the projection represent true to scale the arc distance from the central meridian and this is why the projection is regarded as equidistant (in a limited sense of course). The Cassini–Soldner projection is shown schematically in Figure 4.17. It is noted that $CP = C'P'$; point $C$ is at a higher latitude than $P$ although they have equal $X'$ coordinates; and that point $D$ is at equal latitude to $P$ but its $X'$ coordinate is smaller. The projection is neither conformal nor equivalent.

TABLE 4.1

*Linear distortions*

| Distance from central meridian in km | $M_{max}$ |
|:---:|:---:|
| 0 | 1.000 000 |
| 10 | 1.000 001 |
| 40 | 1.000 020 |
| 70 | 1.000 060 |
| 100 | 1.000 123 |
| 150 | 1.000 277 |
| 200 | 1.000 493 |
| 250 | 1.000 770 |

The mapping equations are given without a proof.

$$X = X_D + \tfrac{1}{2}N(\Delta\lambda \cos \varphi)^2 \tan \varphi + \tfrac{1}{24}N(\Delta\lambda \cos \varphi)^4 \tan \varphi \, (5 - \tan^2 \varphi)$$
$$Y = N(\Delta\lambda \cos \varphi) - \tfrac{1}{6}N(\Delta\lambda \cos \varphi)^3 \tan^2 \varphi -$$
$$\tfrac{1}{120}N(\Delta\lambda \cos \varphi)^5 \tan^2 \varphi \, (8 - \tan^2 \varphi).$$

The maximum linear distortion occurs in the $X$ axis direction and is a function of the distance from the central meridian.

Table 4.1. shows the linear distortion as the projection scale factor.

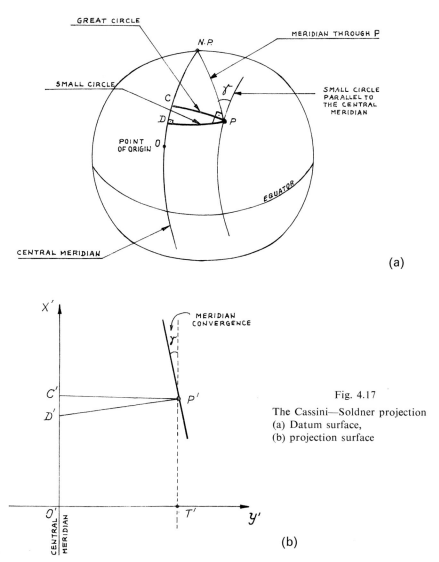

(a)

Fig. 4.17

The Cassini—Soldner projection
(a) Datum surface,
(b) projection surface

(b)

The rapid scale deterioration is clearly visible and this makes the Cassini–Soldner projection unsuitable for areas of considerable extent perpendicular to the central meridian.

## CHAPTER 5

# CONFORMAL PROJECTIONS

## 5.1 General relationships

### 5.1.1 *Introduction*

The general condition for the design of a conformal projection has been laid down in the fundamental transformation matrix (3.21) in combination with the condition (3.24)

$$\frac{E}{e} = \frac{F}{f} = \frac{G}{g} = m^2.$$

This leads to a number of differential equations to which further boundary conditions are added in order to obtain the individual types of transformations. These cannot as a rule be represented geometrically except for one, namely, the stereographic projection of the sphere (see also Section 4.6).

The application of isometric coordinates simplifies the general condition by equalizing the coefficients of the differential quotients. On the ellipsoid this involves the change of the variable $\varphi$ by

$$\psi = \ln \tan \left(45° + \tfrac{1}{2}\varphi\right) + \tfrac{1}{2}\varepsilon \ln \frac{1 - \varepsilon \sin \varphi}{1 + \varepsilon \sin \varphi}$$

as has been shown earlier.

In this section an important common property of conformal projections is discussed: the locus of points of equal scale distortion is a conic section. This is true, neglecting third and higher order terms in the expansions. It will be shown that of the three possibilities the hyperbola is the least important. Projections where these loci are ellipses or parabolae (often degenerated into parallel straight lines) are frequently applied.

### 5.1.2 *The fundamental transformation matrix and the condition of conformality*

The condition of a conformal projection is expressed in formula (3.24) assuming two orthogonal corresponding parametric systems $(F' = F = f = 0)$

$$\frac{E}{e} = \frac{G}{g} = m^2 \tag{5.1}$$

where according to (3.21)

$$
\left.
\begin{array}{l}
E = \left(\dfrac{\partial U}{\partial u}\right)^2 E' + \left(\dfrac{\partial V}{\partial u}\right)^2 G' \\[3mm]
0 = \dfrac{\partial U}{\partial u}\dfrac{\partial U}{\partial v} E' + \dfrac{\partial V}{\partial u}\dfrac{\partial V}{\partial v} G' \\[3mm]
G = \left(\dfrac{\partial U}{\partial v}\right)^2 E' + \left(\dfrac{\partial V}{\partial v}\right)^2 G'
\end{array}
\right\} \tag{5.2}
$$

combination of (5.1) and (5.2) gives the conditions of conformality as

$$
\left.
\begin{array}{l}
E = \left(\dfrac{\partial U}{\partial u}\right)^2 E' + \left(\dfrac{\partial V}{\partial u}\right)^2 G' = em^2 \\[3mm]
0 = \dfrac{\partial U}{\partial u}\dfrac{\partial U}{\partial v} E' + \dfrac{\partial V}{\partial u}\dfrac{\partial V}{\partial v} G' \\[3mm]
G = \left(\dfrac{\partial U}{\partial v}\right)^2 E' + \left(\dfrac{\partial V}{\partial v}\right)^2 G' = gm^2.
\end{array}
\right\} \tag{5.3}
$$

These three relationships cannot be satisfied without imposing further conditions on the differential quotients $\partial U/\partial u$, $\partial U/\partial v$, $\partial V/\partial u$, and $\partial V/\partial v$ or on the transformation formulae (3.3)

$$U = \bar{q}_1(u, v)$$
$$V = \bar{q}_2(u, v) \, .$$

An obvious condition would be that $U = \bar{q}_1(u)$, is a function of $u$ only, so that $\partial U/\partial v = 0$. This implies – see the second equation of (5.3) –

$$\frac{\partial V}{\partial u}\frac{\partial V}{\partial v} G' = 0$$

since $G' \neq 0 \neq \partial V/\partial v$ it follows that $\partial V/\partial u = 0$ and $V$ is a function of $v$ only: $V = \bar{q}_2(v)$.

The equations (5.3) can then be simplified into

$$E = \left(\frac{\partial U}{\partial u}\right)^2 \quad E' = em^2 \left.\begin{array}{l}\\[3em]\\ \end{array}\right\}$$

$$G = \left(\frac{\partial V}{\partial u}\right)^2 \quad G' = gm^2 \qquad (5.4)$$

or

$$\frac{\left(\dfrac{\partial U}{\partial u}\right)^2 E'}{e} = \frac{\left(\dfrac{\partial V}{\partial v}\right)^2 G'}{g} = m^2 \ (= \text{constant}). \qquad (5.5)$$

The two equations (5.5) are integrated separately in order to solve for $U$ and $V$ respectively. In practice $V = \bar{q}_2(v)$ may be selected as simple as possible e.g. a linear function

$$V = c_1 v + c_2 \qquad (5.6)$$

where $c_1$ and $c_2$ are constants. If in addition $G'$ is a function of $U$ only and $g$ of $u$ only, indicated by $G''$ and $g''$ respectively, then

$$\frac{G''}{g''} c_1^2 = m^2$$

and (5.5) becomes

$$\frac{\left(\dfrac{\partial U}{\partial u}\right)^2 E'}{e} = \frac{G''}{g''} c_1^2 \qquad (5.7)$$

whence

$$\sqrt{\frac{E'}{G''}} \, \partial U = c_1 \sqrt{\frac{e}{g''}} \, \partial u. \qquad (5.8)$$

### 5.1.3 *Isometric coordinate systems. Some expansions; the scale distortion*

The units of measure along the coordinate axes $(u, v)$ can be made equal in a manner discussed in Section 3.3.2 (isometric coordinates). The equations (5.3) may then be simplified as to their formulation. It can be argued if this simplification holds computationally especially after the introduction of the electronic computers.

Taking

$$ds^2 = e\,du^2 + g\,dv^2$$

and

$$dS^2 = E'\,dU^2 + G'\,dV^2$$

it is found after transformation:

$$ds^2 = g(d\bar{u}^2 + d\bar{v}^2)$$

with

$$\left.\begin{array}{c} d\bar{u} = \sqrt{\dfrac{e}{g}}\,du \quad \text{or} \quad du = \sqrt{\dfrac{g}{e}}\,d\bar{u} \\[2em] dv = d\bar{v} \end{array}\right\} \tag{5.9}$$

and also

$$dS^2 = G'(d\bar{U}^2 + d\bar{V}^2)$$

with

$$d\bar{U} = \sqrt{\frac{E'}{G'}}\,dU \quad \text{or} \quad dU = \sqrt{\frac{G'}{E'}}\,d\bar{U}. \tag{5.10}$$

By substitution of (5.9) and (5.10) into (5.3) it is seen after some manipulations, that

$$\left(\frac{\partial \bar{U}}{\partial \bar{u}}\right)^2 + \left(\frac{\partial \bar{V}}{\partial u}\right)^2 = \left(\frac{\partial \bar{U}}{\partial v}\right)^2 + \left(\frac{\partial \bar{V}}{\partial \bar{v}}\right)^2 = gm^2 \tag{5.11}$$

and

$$\frac{\partial \bar{U}}{\partial \bar{u}}\frac{\partial \bar{U}}{\partial \bar{v}} = -\frac{\partial \bar{V}}{\partial \bar{u}}\frac{\partial \bar{V}}{\partial \bar{v}}. \tag{5.12}$$

These equations lead to the following relationships, where all coefficients are equalized:

$$\frac{\partial \bar{U}}{\partial \bar{u}} = \pm\frac{\partial \bar{V}}{\partial \bar{v}} \quad \text{and} \quad \frac{\partial \bar{U}}{\partial \bar{v}} = \mp\frac{\partial \bar{V}}{\partial \bar{u}}. \tag{5.13}$$

If there are isometric parameters on both the datum and the image surface, then the projection

$$\bar{U} = q_1(\bar{u},\bar{v}) \quad \text{and} \quad V = q_2(\bar{u},\bar{v}) \tag{5.14}$$

is conformal if the equations (5.13) hold.

Because of the fact that the parameters of analytic functions satisfy the same relationships (5.13) (they are then called the Cauchy–Riemann equations), the partial derivatives in (5.13) can be considered as the partial derivatives of the real and imaginary parts of the equation

$$\bar{U} + i\bar{V} = q(\bar{u} + i\bar{v}) .$$

(5.15)

The real and imaginary parts being

$$q_1(\bar{u}, \bar{v}) \quad \text{and} \quad q_2(\bar{u}, \bar{v})$$

of (5.14) respectively.

The properties of all conformal projections can be described by the complex algebra of analytic functions.

The right hand side of (5.15) may be expanded into a power series:

$$\bar{U} + i\bar{V} = (\bar{u} + i\bar{v}) + (A + iB)(\bar{u} + i\bar{v})^2 + (C + iD)(\bar{u} + i\bar{v})^3 + \dots .$$

(5.16)

It can be corroborated that by the choice of this expansion

(1) The origins in both the date and image plane are corresponding points;
(2) The scale distortion $m$ in the origins is equal to one: $m_0 = 1$.
(3) Both systems of parameters have the same orientation.

Separating the real from the imaginary parts in (5.16) gives

$$\left. \begin{aligned} \bar{U} &= \bar{u} + A(\bar{u}^2 - \bar{v}^2) - 2B\bar{u}\bar{v} + C(\bar{u}^3 - 3\bar{u}\bar{v}^2) + D(-3\bar{u}^2\bar{v} + \bar{v}^3) + \dots \\ \bar{V} &= \bar{v} + A\bar{u}\bar{v} + B(\bar{u}^2 - \bar{v}^2) + C(3\bar{u}^2\bar{v} - \bar{v}^3) + D(\bar{u}^3 - 3\bar{u}\bar{v}^2) + \dots . \end{aligned} \right\}$$

(5.17)

The conformal projections are derived by a suitable choice of the coefficients $A$, $B$, $C$ and $D$. Within the scope of this book, however, this method will not be pursued.

The line elements in the datum and image planes respectively, may be written in terms of isometric parameters

$$ds^2 = \frac{1}{p^2}(d\bar{u}^2 + d\bar{v}^2)$$

and

$$dS^2 = \frac{1}{P^2}(d\bar{U}^2 + d\bar{V}^2).$$

The coefficients $p$ (and $P$) are in general functions of $(u, v)$ $(U, V)$. (See also 3.42).)

The scale distortion in terms of isometric coordinates is

$$m^2 = \frac{p^2(d\bar{U}^2 + dV^2)}{P^2(d\bar{u}^2 + d\bar{v}^2)}$$

(5.18)

since

$$d\overline{U} = \frac{\partial \overline{U}}{\partial \bar{u}} d\bar{u} + \frac{\partial \overline{U}}{\partial \bar{v}} d\bar{v}$$

and

$$d\overline{V} = -\frac{\partial \overline{V}}{\partial \bar{u}} d\bar{u} + \frac{\partial \overline{V}}{\partial \bar{v}} d\bar{v}$$

it is seen that

$$d\overline{U}^2 + d\overline{V}^2 = \left\{ \left(\frac{\partial \overline{U}}{\partial \bar{u}}\right)^2 + \left(\frac{\partial \overline{V}}{\partial \bar{v}}\right)^2 \right\} (d\bar{u}^2 + d\bar{v}^2). \tag{5.19}$$

Substitution of (5.19) into (5.18) leads to

$$m^2 = \frac{p^2}{P^2} \left\{ \left(\frac{\partial \overline{U}}{\partial \bar{u}}\right)^2 + \left(\frac{\partial \overline{U}}{\partial \bar{v}}\right)^2 \right\}. \tag{5.20}$$

Similarly it is derived that

$$m^2 = \frac{p^2}{P^2} \left\{ \left(\frac{\partial \overline{V}}{\partial \bar{u}}\right)^2 + \left(\frac{\partial \overline{V}}{\partial \bar{v}}\right)^2 \right\}. \tag{5.21}$$

Finally attention is given to the relationships obtained by differentiating (5.13). These are

$$\frac{\partial^2 \overline{U}}{\partial \bar{u}^2} + \frac{\partial^2 \overline{U}}{\partial \bar{v}^2} = \frac{\partial^2 \overline{V}}{\partial \bar{u}^2} + \frac{\partial^2 \overline{V}}{\partial \bar{v}^2} = 0. \tag{5.22}$$

They are the Laplace equations for $\overline{U}$ and $\overline{V}$ with respect to the independent parameters $\bar{u}$ and $\bar{v}$.

### 5.1.4   *The loci of points of equal scale distortion*

In this section an important property of all conformal projections is discussed namely "the locus of points of equal scale distortion is by approximation a conic section".

This theorem has first been proven by Tissot [27]. However, the proof may be given in several ways e.g. by the use of isometric coordinates and the expansion of (5.17) collecting the second degree terms; Schols [22] has given an elegant derivation; an extensive treatment is found in [7]. They will not be repeated here.

The ellipsoid is taken as a datum surface with the formula of the elementary distance as given in (3.34)

$$ds^2 = M^2 d\varphi^2 + N^2 \cos^2 \varphi \, d\lambda^2 . \tag{5.23}$$

In the projection plane this distance is given by

$$dS^2 = dX^2 + dY^2 .$$ (5.24)

The parametric curves on the ellipsoid are the meridians and parallel circles of latitude; in the plane a rectangular coordinate system $X$, $Y$ has been adopted. The origins $O$ of both systems are corresponding points with coordinates $\varphi_0$, $\lambda_0$ and $X_0 = Y_0 = 0$ respectively.

Comparing the parametric systems $(U, V)$ and $(u, v)$ (as applied earlier in formulae (5.2) to (5.5) inclusive) with those of (5.23) and (5.24) it is seen that

$$U = X, \quad V = Y; \quad u = \varphi \quad \text{and} \quad v = \lambda .$$

If the projection plane is thought to be tangent to the ellipsoid, the $X$-axis is selected so as to be tangent to the meridian, and the $Y$-axis tangent to the parallel circle at $O$. This results in the first derivatives in the origin being equal to zero:

$$\left(\frac{\partial X}{\partial \lambda}\right)_0 = \left(\frac{\partial Y}{\partial \varphi}\right)_0 = 0.$$ (5.25)

The scale distortion $m_0$ at $O$ is taken as equal to unity. The following expression then, is derived for the scale distortion [22].

$$
m^2 = 1 + \left\{ \frac{1}{M_0 N_0} - \frac{\sin^2 \varphi_0}{N_0^2 \cos^2 \varphi_0} - \frac{1}{N_0^3 \cos^3 \varphi_0} \left(\frac{\partial^3 Y}{\partial \lambda^3}\right)_0 \right\} X^2 +
$$

$$
\left. \begin{array}{c}
- \frac{2}{N_0^3 \cos^3 \varphi_0} \left(\frac{\partial^3 X}{\partial \lambda^3}\right)_0 XY + \\[2mm]
+ \left\{ \frac{\sin^2 \varphi_0}{N_0^2 \cos^2 \varphi_0} + \frac{1}{N_0^3 \cos^3 \varphi_0} \left(\frac{\partial^3 Y}{\partial \lambda^3}\right)_0 \right\} Y^2 .
\end{array} \right\}
$$ (5.26)

Now $\sqrt{MN}$ is an expression known in the differential geometry as the second Gaussian radius of curvature, sometimes called the mean radius of curvature of the ellipsoid. It will occur again when discussing the conformal projection of the ellipsoid to the sphere in Section 5.2.

It may be noted, that because of the occurrence of the term $1/M_0 N_0$, the coefficients of the second degree terms of (5.26) cannot vanish all at the same time for any value of $\varphi$, $\lambda$, or the third derivatives of $X$ and $Y$ with respect to $\lambda$. Denoting

$$A = 1 - \frac{2 M_0 N_0 \sin^2 \varphi_0}{N_0^2 \cos^2 \varphi_0} - \frac{2 M_0 N_0}{N_0^3 \cos^3 \varphi_0} \left(\frac{\partial^3 Y}{\partial^3 \lambda}\right)_0$$ (5.27)

and

$$B = \frac{2 M_0 N_0}{N_0^3 \cos^3 \varphi_0} \left( \frac{\partial^3 X}{\partial \lambda^3} \right)_0$$  (5.28)

(5.26) becomes

$$m^2 = 1 + \frac{(1+A) X^2 - 2 BX Y + (1-A) Y^2}{2 M_0 N_0}$$  (5.29)

or, after the expansion for $\sqrt{1+x} = 1 + \frac{1}{2}x + \ldots$

$$m = 1 + \frac{(1+A) X^2 - 2 BX Y + (1-A) Y^2}{4 M_0 N_0}.$$  (5.30)

This is the equation of a conic section being an ellipse if $A^2 + B^2 < 1$, or a hyperbola if $A^2 + B^2 > 1$. If $A^2 + B^2 = 1$, the curves of equal scale distortion become straight parallel coinciding lines.

These three cases will be elucidated separately. Suppose $A^2 + B^2 < 1$ and suppose that

$$(1+A)X^2 - 2 BX Y + (1-A)Y^2 = C^2$$  (5.31)

is the ellipse which circumscribes as a best fit the circumference of the image. Then

$$m = 1 + \frac{C^2}{4 M_0 N_0}$$  (5.32)

is the maximum scale distortion, radiating from the origin.

By the application of a factor of proportion sometimes called grid scale constant, the scale distortion $m_0$ at $O$ may be made equal to

$$m_0 = 1 - \frac{C^2}{8 M_0 N_0},$$  (5.33)

the scale distortion at the outer circumference then becoming

$$m_c = 1 + \frac{C^2}{8 M_0 N_0}.$$  (5.34)

This causes the maximum deviation from $m = 1$ to be equal to $C^2/8 M_0 N_0$ instead of $C^2/4 M_0 N_0$.

In some projections this procedure corresponds geometrically to a change of position of the projection plane parallel to itself so that it intersects the datum surface instead of being tangent at $O$ (see Fig. 5.1).

Consider the values of $A$ and $B$ in (5.31) rendering $C^2$ a minimum. In Figure 5.2 the ellipse of (5.30) is shown.

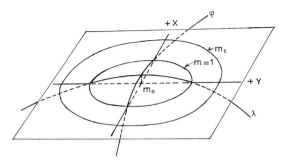

Fig. 5.1
Ellipse of equal scale distortion

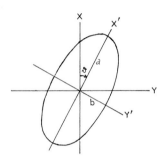

Fig. 5.2
Application of grid scale constant

The semi major and minor axes $a$ and $b$ coincide with the $X'$ and $Y'$ axes, a coordinate system obtained by rotation of the $X$, $Y$ system about an angle $\vartheta$. The equation of the ellipse on the $X'Y'$ system is

$$\frac{X'^2}{a^2} + \frac{Y'^2}{b^2} = 1 \tag{5.35}$$

the rotation being given by

$$\begin{vmatrix} X' \\ Y' \end{vmatrix} = \begin{vmatrix} \cos \vartheta & \sin \vartheta \\ -\sin \vartheta & \cos \vartheta \end{vmatrix} \begin{vmatrix} X \\ Y \end{vmatrix}.$$

The equation of (5.31) on the $X$, $Y$ system is

$$\left.\begin{aligned} &\left(\frac{\cos^2 \vartheta}{a^2} + \frac{\sin^2 \vartheta}{b^2}\right)X^2 + 2\left(\frac{1}{a^2} - \frac{1}{b^2}\right)\sin \vartheta \cos \vartheta\, XY + \\ &+ \left(\frac{\sin^2 \vartheta}{a^2} + \frac{\cos^2 \vartheta}{b^2}\right)Y^2 = 1 \end{aligned}\right\} \tag{5.36}$$

This ellipse should be identical to (5.30), whence

$$
\left.
\begin{aligned}
(1+A) &= C^2\left(\frac{\cos^2 \vartheta}{a^2} + \frac{\sin^2 \vartheta}{b^2}\right) \\[2ex]
B &= C^2\left(\frac{1}{a^2} - \frac{1}{b^2}\right)\sin^2 \vartheta \\[2ex]
(1-A) &= C^2\left(\frac{\sin^2 \vartheta}{a^2} + \frac{\cos^2 \vartheta}{b^2}\right).
\end{aligned}
\right\}
\tag{5.37}
$$

These are three equations with three unknowns $A$, $B$ and $C^2$. The solution is

$$
\left.
\begin{aligned}
C^2 &= \frac{2a^2 b^2}{a^2 + b^2} \\[2ex]
A &= \frac{a^2 - b^2}{a^2 + b^2}\cos 2\vartheta \\[2ex]
B &= \frac{a^2 - b^2}{a^2 + b^2}\sin 2\vartheta.
\end{aligned}
\right\}
\tag{5.38}
$$

Now $C$ appears to be the length of the distance between the origin and a point with coordinates on the ellipse $X' = Y' = \tfrac{1}{2}C\sqrt{2}$ on the bisectrix of the angle $X'OY'$.

Thus the problem consists of finding an ellipse the circumference of which is a best fit of the circumference of the mapped area. In addition the length of the bisectrix of the intersecting angle of the principal axes ($2C$) should be minimum.

Together with the coordinates $\varphi_0$ and $\lambda_0$ of the origin there are 5 unknowns. Hence the coordinates of 5 points on the circumference suffice to calculate the elements of the ellipse. Redundant points give cause for a statistical problem of curve fitting.

Considering the formulae (5.27) and (5.28) for $A$ and $B$ respectively, it is noted that the values of (5.38) rendering $C^2$ a minimum impose conditions on the derivatives $(\partial^3 X/\partial X^3)_0$ and $(\partial^3 Y/\partial \lambda^3)_0$, and thus on the special properties of the transformation. This is not always possible.

The ellipse becomes a circle if $A = 0$.

The case $A^2 + B^2 = 1$ makes it possible that the expression in (5.30)

$$
(1+A)X^2 - 2BXY + (1-A)Y^2
$$

be factorized as

$$\left( X - \frac{B}{1+A} Y \right)^2 \tag{5.39}$$

yielding

$$m = 1 + \frac{\left( X - \dfrac{B}{1+A} \right)^2}{4 M_0 N_0}. \tag{5.40}$$

The locus of points where $m = 1$ is indicated by the line

$$X - \frac{B}{1+A} Y = 0. \tag{5.41}$$

An investigation of the gradient shows that it coincides with the $X'$ axis of (5.35).

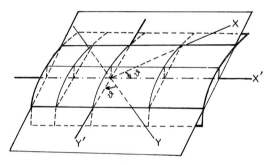

Fig. 5.3
Application of grid scale constant

In this case, parallel lines of equal scale distortion are obtained (see Fig. 5.3). Comparison is made with the ellipse taking $a = \infty$. The quantities in (5.38) then are

$$C^2 = 2 b^2; \quad A = + \cos 2\vartheta; \quad B = + \sin 2\vartheta \tag{5.42}$$

The maximum variation from unity is

$$\frac{C^2}{8 M_0 N_0} = \frac{b^2}{4 M_0 N_0} = \frac{1}{16} \left( \frac{2b}{\sqrt{M_0 N_0}} \right)^2. \tag{5.43}$$

The distance between the lines of this variation is $2b$.

If $(\partial^3 X / \partial \lambda^3)_0 = 0$, then $B = 0$ in (5.28), and the axes $X'Y'$ coincide with the axes of the $X$, $Y$ system.

Putting $\vartheta = 0$ in (5.38) it is seen that

$$A = \frac{a^2 - b^2}{a^2 + b^2}.$$

(5.44)

For $B = 0$ and $A = +1$ (5.30) becomes a parabola.

Points of equal scale distortion are situated on hyperbolae if $A^2 + B^2 > 1$. This case is less favourable than the two previous ones. The two asymptotic lines intersecting in the origin are the loci where $m = 1$. The scale distortion increases with one set of hyperbolae but decreases with the conjugate set (Fig. 5.4).

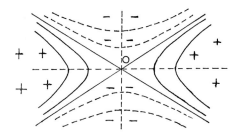

Fig. 5.4
Application of grid scale constant

If in one case

$$(1 + A)X^2 - 2BXY + (1 - A)Y^2 = C_1^2$$

and in the other

$$(1 + A)X^2 - 2BXY + (1 - A)Y^2 = C_2^2$$

then the maximum variation is equal to

$$\frac{C_1^2 + C_2^2}{8M_0N_0}$$

(5.45)

which usually does not improve on the previous elliptic and parabolic types of projections. The hyperbolic type of conformal projection is therefore considered of less importance.

## 5.2   The conformal projection of the ellipsoid on the sphere

### 5.2.1   *Introduction*

The theorem that a successive number of conformal transformations yields a conformal image of the original figure, appears to be of advantage to the conformal mapping of the earth on the plane.

It is possible to project the ellipsoid conformally to a body of an equal mean curvature $1/\sqrt{M_0 N_0}$ in the origin, and then on to the plane. Gauss made use of this property by projecting the ellipsoid to the sphere directly. Hotine [11] used the "aposphere" as an intermediate projection surface between these two surfaces. Both methods give manageable formulae. However, in this section only the former method will be discussed referring to the literature for the latter method.

It should be realised that the indirect or double projection through the sphere is not exactly the same as the direct one in that they differ concerning the higher order terms of the expanded transformation formulae.

Once the dimensions of the sphere as a projection surface have been determined, any great circle may act as a zero meridian, the great circle at right angles to it acting as an equator or zero parallel. Thus a new system of "longitude" curves and "parallel" circles can be adopted, using pure spherical trigonometry for calculation. This will facilitate the formulation of oblique projections.

It will be derived that the sphere has a radius of $R = \sqrt{M_0 N_0}$ which is the mean radius of curvature at the central point on the ellipsoid, and that the transformation formulae are

$$\tan\left(45° + \tfrac{1}{2}\Phi\right) = \left\{\tan\left(45° + \tfrac{1}{2}\varphi\right)\left(\frac{1 - \varepsilon \sin \varphi}{1 + \varepsilon \sin \varphi}\right)^{\frac{1}{2}\varepsilon}\right\}^c$$

and

$$\Lambda = c\lambda \tag{5.46}$$

where

$$c = (1 + \varepsilon'^2 \cos^4 \varphi_0)^{\frac{1}{2}} = \left\{1 + \frac{\varepsilon^2 \cos^4 \varphi_0}{1 - \varepsilon^2}\right\}^{\frac{1}{2}}.$$

This relatively simple result is due to the fact that the second derivative of the scale distortion $(\partial^2 m^2/\partial \varphi^2)_0$ in the origin can be made equal to zero. It has been shown in the previous section that this cannot be done when transforming the ellipsoid into a plane directly.

### 5.2.2   *The transformation formula; the radius of the conformal sphere*

The elementary distance on the ellipsoid has been given previously as

$$ds^2 = M^2 \, d\varphi^2 + N^2 \cos^2 \varphi \, d\lambda^2 . \tag{5.47}$$

with the fundamental quantities

$$e = M^2 \quad \text{and} \quad g = N^2 \cos^2 \varphi . \tag{5.48}$$

The elementary distance on the sphere (in this case a projection surface, so capitals are being used)

$$dS^2 = R^2 \, d\Phi^2 + R^2 \cos^2 \Phi d\Lambda^2 \tag{5.49}$$

with

$$E' = R^2 \quad \text{and} \quad G' = R^2 \cos^2 \Phi . \tag{5.50}$$

As transformation conditions are laid down: the spherical latitude $\Phi$ is a function of the ellipsoidal geodetic latitude $\varphi$ only; the spherical longitude $\Lambda$ is a linear function of the ellipsoidal longitude viz.

$$\left.\begin{aligned} \Phi &= f(\varphi) \\ \Lambda &= c\lambda + c_1 . \end{aligned}\right\} \tag{5.51}$$

whence $\partial \Lambda / \partial \lambda = c$ ($c$ and $c_1$ are constants).

The fundamental transformation matrix (3.21) then gives

$$\left.\begin{aligned} E &= \left(\frac{\partial \Phi}{\partial \varphi}\right)^2 R^2 \\ G &= c^2 R^2 \cos^2 \varphi. \end{aligned}\right\} \tag{5.52}$$

The condition of conformality is expressed by (5.5)

$$\frac{1}{M^2}\left(\frac{\partial \Phi}{\partial \varphi}\right)^2 R^2 = \frac{c^2 R^2 \cos^2 \Phi}{N^2 \cos^2 \varphi} = m^2 \tag{5.53}$$

whence

$$\frac{\partial \Phi}{\cos \Phi} = \frac{cM}{N \cos \varphi} \, d\varphi. \tag{5.54}$$

The solution of this differential equation is

$$\ln \tan (45^\circ + \tfrac{1}{2}\Phi) = c \ln \tan (45^\circ + \tfrac{1}{2}\varphi)\left(\frac{1 - \varepsilon \sin \varphi}{1 + \varepsilon \sin \varphi}\right)^{\frac{1}{2}\varepsilon} \tag{5.55}$$

or

$$\tan (45^\circ + \tfrac{1}{2}\Phi) = \left\{\tan (45^\circ + \tfrac{1}{2}\varphi)\left(\frac{1 - \varepsilon \sin \varphi}{1 + \varepsilon \sin \varphi}\right)^{\frac{1}{2}\varepsilon}\right\}^c . \tag{5.56}$$

The reader is asked to compare this result by combining the formulae (3.37) and (3.39) using isometric coordinates.

Since it is wanted that in the origin

$$\Lambda_0 = c\lambda_0 \tag{5.57}$$

the known coefficient $c_1$ is taken as equal to zero.

The scale distortion $m$ at a point $P$ is given as a Taylor expansion, noting, that the derivatives to $\lambda$ are equal to zero:

$$m^2 = m_0^2 + \left(\frac{\partial m^2}{\partial \varphi}\right)_0 \Delta\varphi + \frac{1}{2}\left(\frac{\partial^2 m^2}{\partial \varphi^2}\right)_0 \Delta\varphi^2 + \dots \qquad (5.58)$$

Also, as has been done previously, the scale distortion in the origin $O$ will be assumed to be equal to unity. It will now be shown, that the first and second order terms in (5.58), may vanish, under certain conditions.

From (5.53)

$$m = \frac{cR \cos \Phi}{N \cos \varphi} \qquad (5.59)$$

and

$$m_0 = 1 = \frac{cR \cos \Phi_0}{N_0 \cos \varphi_0} . \qquad (5.60)$$

Now

$$\left(\frac{\partial m}{\partial \varphi}\right)_0 = cR \left\{\frac{\partial \cos \Phi \, (N \cos \varphi)^{-1}}{\partial \varphi}\right\}_0 = 0$$

or

$$\sin \Phi_0 \left(\frac{\partial \Phi}{\partial \varphi}\right)_0 = \cos \Phi_0 \frac{M_0 \sin \varphi_0}{N_0^2 \cos^2 \varphi_0}$$

which results in, by the use of (5.53)

$$\sin \varphi_0 = c \sin \Phi_0 . \qquad (5.61)$$

Differentiation of $(\partial m / \partial \varphi)$, putting $(\partial^2 m^2 / \partial \varphi^2)_0 = 0$, and equation (5.61) lead to

$$\tan \varphi_0 = \sqrt{\frac{N_0}{M_0}} \tan \Phi_0 . \qquad (5.62)$$

Solution of $c$ from (5.61) and (5.62) by elimination of the terms with $\Phi_0$ gives

$$c = \left(1 + \frac{\varepsilon^2 \cos^4 \varphi_0}{1 - \varepsilon^2}\right)^{\frac{1}{2}} \qquad (5.63)$$

or using

$$\varepsilon'^2 = \frac{a^2 - b^2}{b^2},$$

$$c = (1 + \varepsilon'^2 \cos^4 \varphi_0)^{\frac{1}{2}}. \qquad (5.64)$$

By the substitution of

$$c = \frac{\cos \varphi_0}{\cos \Phi_0} \sqrt{\frac{N_0}{M_0}},$$

(which also follows from (5.61) and (5.62)) into (5.60)

$$m_0 = \frac{cR \cos \Phi_0}{N_0 \cos \varphi_0}, \qquad (5.65)$$

it is shown that

$$m_0 = \frac{R}{\sqrt{M_0 N_0}}. \qquad (5.66)$$

In the central point $m_0 = 1$, whence

$$R = \sqrt{M_0 N_0}. \qquad (5.67)$$

## 5.3  Projections on the cone, the cylinder and the plane

### 5.3.1  *Introduction*

A general derivation of three allied types of conformal projections of the ellipsoid on a cone, a cylinder and a plane will be developed in this section. They are the Lambert normal conical projections (Section 5.3.2) with one and two standard parallels; the Mercator projection on a cylindrical surface, tangent at

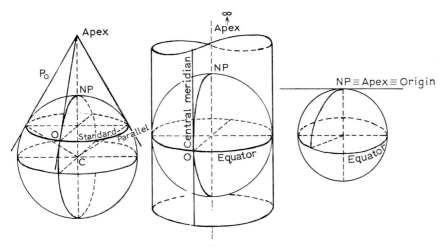

Fig. 5.5
Normal conical, cylindrical and plane projections

the equator; and the polar stereographic projection (Section 5.3.3), where the central point coincides with the (North) pole. The three cases are schematically shown in Figure 5.5. The Mercator and the stereographic projection are considered to originate from the conical projection by a shift of the Apex $A$ of the cone to infinity, and to the pole respectively. The corresponding transformation formulae for the sphere follow by putting the eccentricity $\varepsilon$ equal to zero.

### 5.3.2 The Lambert conical projection. The scale distortion. Two standard parallels

In the case of the Lambert conformal projection with one standard parallel the conical surface is tangent to the ellipsoid at the standard parallel; the apex $A$ of the cone lies on the rotational North–South pole axis of the ellipsoid. It has already been mentioned that, after developing the conical surface, the meridians are straight lines converging at $A$, this point also being the centre of all projected circular parallels. One meridian is adopted as the central or zero meridian (Fig. 5.6), the intersection of this line with the standard parallel being the central point or origin $O$, with ellipsoidal $\varphi_0, \lambda_0$.

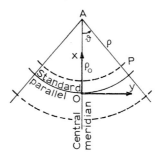

Fig. 5.6
Development of conical projection surface

The coordinates of an arbitrary point $P$ are given in a polar coordinate system $(\rho, \vartheta)$:

$$\rho = AP = \rho_P \quad \text{and} \quad \vartheta = \vartheta_P \tag{5.68}$$

The distance $AO = \rho_0$. Adopting an $X, Y$ rectangular coordinate system with the abscissa along the zero meridian, and the $Y$ at right angles to it, the relationship between the two systems is

$$\left.\begin{array}{l} X = \rho_0 - \rho \cos \vartheta \\ Y = \rho \sin \vartheta . \end{array}\right\} \tag{5.69}$$

For the derivations on the ellipsoid the line element is taken again as

$$ds^2 = M^2 \, d\varphi^2 + N^2 \cos^2 \varphi \, d\lambda^2 \tag{5.70}$$

with

$$e = M^2 \quad \text{and} \quad g = N^2 \cos^2 \varphi.$$

On the conical surface this element is given by (3.40), taking the pole of the system at the apex $A$:

$$dS^2 = d\rho^2 + \rho^2 \, d\vartheta^2 \tag{5.71}$$

with

$$E' = 1 \quad \text{and} \quad G' = \rho^2 .$$

Conditions are that

(1) $\rho$ is a function of $\varphi$ only:

$$\rho = q_1(\varphi) . \tag{5.72}$$

and

(2) $\vartheta$ is a linear function of $\lambda$

$$\vartheta = c_1 \lambda + c_2 . \tag{5.73}$$

Since $d\vartheta/d\lambda = c_1$, (3.21) gives

$$\left.\begin{aligned} E &= \left(\frac{\partial \rho}{\partial \varphi}\right)^2 \\[2ex] G &= \left(\frac{\partial \vartheta}{\partial \lambda}\right)^2 \rho^2 = c_1^2 \rho^2 . \end{aligned}\right\} \tag{5.74}$$

The conditions of conformality are

$$\frac{1}{M^2}\left(\frac{\partial \rho}{\partial \varphi}\right)^2 = \frac{c_1^2 \rho^2}{N^2 \cos^2 \varphi} = m^2 \tag{5.75}$$

whence, noting that $\rho$ decreases with an increment of $\varphi$

$$\frac{\partial \rho}{\rho} = - \frac{Mc_1}{N \cos \varphi} \, d\varphi. \tag{5.76}$$

Integration yields

$$\ln \rho = -c_1 \ln \left\{ \tan (45° + \tfrac{1}{2}\varphi) \left(\frac{1 - \varepsilon \sin \varphi}{1 + \varepsilon \sin \varphi}\right)^{\frac{1}{2}\varepsilon} + \ln c_3 \right\} \tag{5.77}$$

(see also (3.66)),
so that

$$\rho = c_3 \left\{ \tan (45° - \tfrac{1}{2}\varphi) \left( \frac{1 + \varepsilon \sin \varphi}{1 - \varepsilon \sin \varphi} \right)^{\frac{1}{2}\varepsilon} \right\}^{c_1} . \tag{5.78}$$

The constants $c_1$, $c_2$ and $c_3$ will now be determined; first the $c_3$: in the origin $O$ $(\varphi_0, \lambda_0)$ the cone being tangent to the ellipsoid:

$$\rho = \rho_0 = N_0 \cot \varphi_0 . \tag{5.79}$$

Using (5.78) it is seen that

$$\rho_0 = N_0 \cot \varphi_0 = c_3 \left\{ \tan (45° - \tfrac{1}{2}\varphi_0) \left( \frac{1 + \varepsilon \sin \varphi_0}{1 - \varepsilon \sin \varphi_0} \right)^{\frac{1}{2}\varepsilon} \right\}^{c_1} .$$

Therefore

$$c_3 = \frac{N_0 \cot \varphi_0}{\left\{ \tan (45° - \tfrac{1}{2}\varphi_0) \left( \dfrac{1 + \varepsilon \sin \varphi_0}{1 - \varepsilon \sin \varphi_0} \right)^{\frac{1}{2}\varepsilon} \right\}^{c_1}} . \tag{5.80}$$

The scale distortion $m$ is according to (5.75)

$$m = \frac{c_1 \rho}{N \cos \varphi} . \tag{5.81}$$

As $m_0 = 1$, and the variation of $m$ should be minimum radiating from the origin, the condition should be satisfied that $(\partial m / \partial \varphi)_0$ is equal to zero. Differentiating (5.81) with respect to $\varphi$ gives at the origin:

$$\left( \frac{\partial m}{\partial \varphi} \right)_0 = 0 = c_1 \left( \frac{\partial \rho}{\partial \varphi} \right)_0 (N_0 \cos \varphi_0)^{-1} + c_1 \rho_0 \frac{M_0 \sin \varphi_0}{N_0^2 \cos^2 \varphi_0} . \tag{5.82}$$

Now

$$\left( \frac{\partial \rho}{\partial \varphi} \right)_0 = - \frac{c_1 \rho_0 M_0}{N_0 \cos \varphi_0} = - m_0 M_0$$

since $\rho_0 = N_0 \cot \varphi_0$.

Hence (5.82) becomes, since $m_0 = 1$ (see Section 5.1.4)

$$0 = - \rho_0 M_0 c_1 + \rho_0 M_0 \sin \varphi_0 . \tag{5.83}$$

or

$$c_1 = \sin \varphi_0$$

which is in fact the constant of the cone. The constant $c_2$ is equal to zero, if at the origin

$$\vartheta = 0 \quad \text{if} \quad \lambda = 0 .$$

The transformation formulae of this Lambert conical projection thus become

$$\rho = \rho_0 \left\{ \frac{\tan\left(45° - \tfrac{1}{2}\varphi\right) \left(\dfrac{1 + \varepsilon \sin \varphi}{1 - \varepsilon \sin \varphi}\right)^{\tfrac{1}{2}\varepsilon}}{\tan\left(45° - \tfrac{1}{2}\varphi_0\right) \left(\dfrac{1 + \varepsilon \sin \varphi_0}{1 - \varepsilon \sin \varphi_0}\right)^{\tfrac{1}{2}\varepsilon}} \right\}^{\sin \varphi_0}$$

$$\vartheta = \lambda \sin \varphi_0$$

$$m = \frac{\rho \sin \varphi_0}{N \cos \varphi}$$

$$\left.\begin{array}{r}\\[6pt] \\[6pt] \\[6pt] \\[6pt] \\[6pt] \\[6pt]\end{array}\right\} . \qquad (5.84)$$

It may be noted that the scale distortion is dependent only on the latitude $\varphi$, and not on the longitude $\lambda$, therefore the scale distortion at a parallel circle is constant making the projection suitable for areas extended in an East–West direction.

Prior to a discussion of the Mercator and the polar stereographic projection which follow from (5.84) by putting $\varphi_0 = 0°$ (sin $\varphi_0 = 0$) and $\varphi_0 = 90°$ (sin $\varphi_0 = 1$) respectively, the scale distortion will be treated in more detail.

The scale distortion may be given by an expansion of $\varphi$ only (see (5.78) and (5.81)).

$$m = 1 + \frac{1}{2}\left(\frac{\partial^2 m}{\partial \varphi^2}\right)_0 \Delta\varphi^2 . \qquad (5.85)$$

This is valid for a narrow area stretching along a parallel circle.

By differentiation of (5.81) twice it is found that

$$\left(\frac{\partial^2 m}{\partial \varphi^2}\right)_0 = \frac{M_0}{N_0} \qquad (5.86)$$

whence

$$m = 1 + \frac{M_0}{2 N_0} \Delta\varphi^2 . \qquad (5.87)$$

Formula (5.87) is independent of the quantity $c_1 = \sin \varphi_0$, making it valid for both the Lambert and the Mercator projection. This is *not* so for the stereographic projection because of the fact that

$$\lim_{\varphi_0 \to 90°} \left(\frac{\partial^2 m}{\partial \varphi_0^2}\right) \neq \frac{M_0}{N_0} \quad \text{or} \quad \lim_{\varphi_0 \to 90°} \left(\frac{\partial^2 m}{\partial \varphi_0^2}\right) \neq 1 .$$

Using the first order terms only,

$$(\rho - \rho_0)^2 = \varDelta \rho^2 = \left(\frac{\partial \rho}{\partial \varphi}\right)_0^2 \varDelta \varphi^2.$$

Since

$$\left.\begin{array}{l} \left(\dfrac{\partial \rho}{\partial \varphi}\right)_0 = -M_0 \\[4mm] \varDelta \varphi^2 = \dfrac{\varDelta \rho^2}{M^2} \end{array}\right\} \tag{5.88}$$

whence

$$m = 1 + \frac{1}{2 M_0 N_0} \varDelta \rho^2. \tag{5.89}$$

By expansion of (5.69):

$$X = (\rho_0 - \rho) - \rho \frac{\vartheta^2}{2!}$$

$$Y = \rho \vartheta .$$

Hence to a sufficient approximation

$$m = 1 + \frac{X^2}{2 M_0 N_0}. \tag{5.90}$$

Thus it is seen that the Lambert conical projection is a parabolic type of projection. It is shown below that this holds for the Mercator projection also.

By changing the position of the projection cone the maximum variation of the scale distortion was according to (5.43)

$$\frac{c^2}{8 M_0 N_0} = \frac{\varDelta \rho^2}{4 M_0 N_0} = \frac{1}{16}\left(\frac{2\varDelta \rho}{\sqrt{M_0 N_0}}\right)^2$$

The distance between the two parallel circles is equal to $2\varDelta \rho$ and the limit of the scale distortion is

$$1 \pm \frac{1}{16}\left(\frac{2\varDelta \rho}{\sqrt{M_0 N_0}}\right)^2$$

The area is fitted as close as possible between two parallel circles. For less narrow areas, however, the rigorous formulae must be used.

Suppose the area to be mapped fits between the parallel circles of latitude $\varphi_1$ and $\varphi_2$, and that at these limits the scale distortion should be the same.

Then

$$m_1 = \frac{\rho_1 \sin \varphi_0}{N_1 \cos \varphi_1} = m_2 = \frac{\rho_2 \sin \varphi_0}{N_2 \cos \varphi_0} \qquad (5.91)$$

or

$$\frac{\rho_1}{\rho_2} = \frac{N_1 \cos \varphi_1}{N_2 \cos \varphi_2}$$

or

$$\frac{N_1 \cos \varphi_1}{N_2 \cos \varphi_2} = \frac{\left\{ \tan (45° - \tfrac{1}{2}\varphi_1) \left( \dfrac{1+\varepsilon \sin \varphi_1}{1-\varepsilon \sin \varphi_1} \right)^{\frac{1}{2}\varepsilon} \right\}^{\sin \varphi_0}}{\left\{ \tan (45° - \tfrac{1}{2}\varphi_2) \left( \dfrac{1+\varepsilon \sin \varphi_2}{1-\varepsilon \sin \varphi_2} \right)^{\frac{1}{2}\varepsilon} \right\}^{\sin \varphi_0}} . \qquad (5.92)$$

This expression determines the value of $\sin \varphi_0$, and therefore the latitude $\varphi_0$ of the central parallel circle:

$$\sin \varphi_0 = \frac{\ln N_1 \cos \varphi_1 - \ln N_2 \cos \varphi_2}{\ln \tan (45° - \tfrac{1}{2}\varphi_1)\left( \dfrac{1+\varepsilon \sin \varphi_1}{1-\varepsilon \sin \varphi_1} \right)^{\frac{1}{2}\varepsilon} - \ln \tan (45° - \tfrac{1}{2}\varphi_2)\left( \dfrac{1+\varepsilon \sin \varphi_2}{1-\varepsilon \sin \varphi_2} \right)^{\frac{1}{2}\varepsilon}} .$$

$$(5.93)$$

In this case the conical surface is tangent at the central parallel circle, where the scale distortion $m_0 = 1$.

One may now fix the scale distortion equal to unity at two parallel circles $\varphi_1$ and $\varphi_2$ selected a priori. From (5.91) follows

$$\frac{\rho_{\varphi_1} \sin \varphi_0}{N_{\varphi_1} \cos \varphi_1} = \frac{\rho_{\varphi_2} \sin \varphi_0}{N_{\varphi_2} \cos \varphi_2} = 1$$

or

and

$$\left. \begin{array}{l} \rho_{\varphi_1} \sin \varphi_0 = N_{\varphi_1} \cos \varphi_1 \\[2mm] \rho_{\varphi_2} \sin \varphi_0 = N_{\varphi_2} \cos \varphi_2 . \end{array} \right\} \qquad (5.94)$$

The $\sin \varphi_0$ remains as given by (5.93). Having substituted

$$\rho_{\varphi_1} = \rho_0 \left\{ \frac{\tan (45° - \tfrac{1}{2}\varphi_1)\left( \dfrac{1+\varepsilon \sin \varphi_1}{1-\varepsilon \sin \varphi_1} \right)^{\frac{1}{2}\varepsilon}}{\tan (45° - \tfrac{1}{2}\varphi_0)\left( \dfrac{1+\varepsilon \sin \varphi_0}{1-\varepsilon \sin \varphi_0} \right)^{\frac{1}{2}\varepsilon}} \right\}^{\sin \varphi_0}$$

into (5.94), a second constant $C$ for this projection may be solved

$$
\left.
\begin{aligned}
C &= \frac{\rho_0}{\tan(45° - \frac{1}{2}\varphi_0)\left(\dfrac{1+\varepsilon \sin \varphi_0}{1-\varepsilon \sin \varphi_0}\right)^{\frac{1}{2}\varepsilon}} = \\[2ex]
&= \frac{N_{\varphi_1} \cos \varphi_1}{\sin \varphi_0 \left\{\tan(45° - \frac{1}{2}\varphi_1)\left(\dfrac{1+\varepsilon \sin \varphi_1}{1-\varepsilon \sin \varphi_1}\right)^{\frac{1}{2}\varepsilon}\right\}^{\sin \varphi_0}} \\[2ex]
&= \frac{N_{\varphi_2} \cos \varphi_2}{\sin \varphi_0 \left\{\tan(45° - \frac{1}{2}\varphi_2)\left(\dfrac{1+\varepsilon \sin \varphi_2}{1-\varepsilon \sin \varphi_2}\right)^{\frac{1}{2}\varepsilon}\right\}^{\sin \varphi_0}}
\end{aligned}
\right\} \qquad (5.95)
$$

This is the Lambert conical conformal projection with two standard parallels $\varphi_1$ and $\varphi_2$. The cone intersects the ellipsoid at these parallel circles (Fig. 5.7).

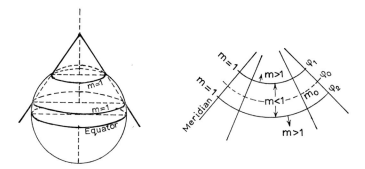

Fig. 5.7
Two standard parallels

The equations thus become

$$
\rho = C \left\{\tan(45° - \frac{1}{2}\varphi)\left(\frac{1+\varepsilon \sin \varphi}{1-\varepsilon \sin \varphi}\right)^{\frac{1}{2}\varepsilon}\right\}^{\sin \varphi_0}
$$

$$
\theta = \lambda \sin \varphi_0
$$

$$
m = \frac{\rho \sin \varphi_0}{N \cos \varphi}.
$$

The scale distortions are illustrated by the drawn curve in Figure 5.8 for the cases $\varphi_1 = 45°$ and $\varphi_2 = 33°$ (see [2]). The value of $\varphi_0$ is equal to $39°05'13''.27$.

As a comparison the dotted line in this figure gives the corresponding scale distortions for the projection with one standard parallel.

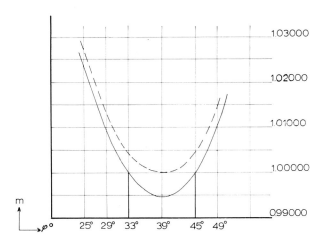

Fig. 5.8
Scale distortions

Another type of condition may be formulated namely that the product of the maximum scale distortion $m_1$ at the parallel circles limiting the area and the minimum scale distortion $m_0$ at the central parallel circle should be equal to unity:

$$m_1 \, m_0 = m_2 \, m_0 = 1 \, .$$

The calculation of the constant in (5.95) is left to the reader.

### 5.3.3  *The Mercator and polar stereographic projections. The scale distortion*

The cylinder of the Mercator projection cut open, shows the meridians as straight parallel lines with, at right angles, the parallel circles also as straight lines. Since $\rho_0 = \infty$ and $\vartheta = 0$ the meridians are equally spaced. It is therefore more convenient to use the $X$, $Y$ system in the calculations directly instead of the polar coordinates $(\rho, \vartheta)$. The intervals between the parallels increase proportionally to sec $\varphi$, as will be shown. The equator acts as the $Y$ axis. The grid of meridians and parallels is shown in Figure 5.9.

If the central point is on the equator – which is the case in the Mercator projection $(\varphi_0 = \sin \varphi_0 = 0)$ –, then

$$\rho_0 = (N_0 \cot \varphi_0)_{\varphi_0 = 0} = \infty$$

and

$$\vartheta = 0 \, .$$

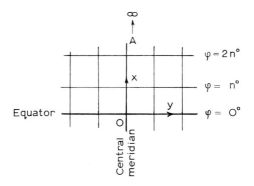

Fig. 5.9
Grid of meridians and parallels (Mercator)

The standard parallel is the equator portrayed as a straight line. In order to derive the transformation formulae the variables $\rho$ and $\vartheta$ must be changed. From (5.69) it is derived that

$$\left.\begin{aligned} dX &= -\cos\vartheta\,d\rho + \rho\sin\vartheta\,d\vartheta \\ dY &= \sin\vartheta\,d\rho + \rho\cos\vartheta\,d\vartheta\,. \end{aligned}\right\} \tag{5.96}$$

Then (5.71) becomes

$$dS^2 = dX^2 + dY^2\,. \tag{5.97}$$

For $\vartheta = 0$

$$\left.\begin{aligned} dX &= -d\rho \\ dY &= \rho\,d\vartheta\,. \end{aligned}\right\} \tag{5.98}$$

So that

$$\frac{\partial\rho}{\partial\varphi} = -\frac{\partial X}{\partial\varphi} \quad\text{and}\quad \frac{\partial\vartheta}{\partial\lambda} = \frac{1}{\rho}\frac{\partial Y}{\partial\lambda}\,. \tag{5.99}$$

After substitution of these expressions into (5.75), (5.76) and (5.82) it is easily corroborated that the transformation formulae of the Mercator projection are

$$\left.\begin{aligned} X &= a\ln\tan(45° + \tfrac{1}{2}\varphi)\left(\frac{1-\varepsilon\sin\varphi}{1+\varepsilon\sin\varphi}\right)^{\frac{1}{2}\varepsilon} \\[2mm] Y &= a\lambda \end{aligned}\right\} \tag{5.100}$$

where $a$ is the radius of the equatorial circle.

The scale distortion is, combining (5.89) and (5.98):

$$m = \frac{a}{N\cos\varphi} = 1 + \frac{X^2}{2M_0N_0} = 1 + \frac{(1-\varepsilon^2)X^2}{2a^2} \tag{5.101}$$

The meridians in polar stereographic projection are straight lines converging at the pole – i.e. the central point. The projected parallels are concentric circles about this point. Taking the $X$-axis along the central meridian, and the $Y$-axis along the meridian perpendicular to it, it is seen that the conversion between the polar and rectangular systems is given by

$$\left. \begin{array}{l} X = \rho \cos \vartheta \\ Y = \rho \sin \vartheta . \end{array} \right\} \tag{5.102}$$

An illustration is given in Figure 5.10.

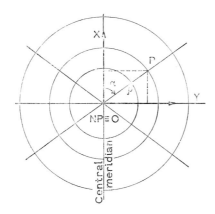

Fig. 5.10
Meridians and parallels (Polar stereographic)

The polar stereographic projection is obtained by putting $\varphi_0 = 90°$ and $\sin \varphi_0 = 1$. Substituting this into (5.84):

$$\rho = \tan (45° - \tfrac{1}{2} \varphi) \left( \frac{1 + \varepsilon \sin \varphi}{1 - \varepsilon \sin \varphi} \right)^{\frac{1}{2}\varepsilon} \frac{a}{(1 - \varepsilon^2)^{\frac{1}{2}}} \lim_{\varphi_0 \to 90°} \cot \varphi_0 \tan^{-1} (45° - \tfrac{1}{2}\varphi_0)$$

or

$$\rho = \tan (45° - \tfrac{1}{2} \varphi) \left( \frac{1 + \varepsilon \sin \varphi}{1 - \varepsilon \sin \varphi} \right)^{\frac{1}{2}\varepsilon} \frac{a}{(1 - \varepsilon^2)^{\frac{1}{2}}} \lim_{\varphi_0 \to 90°} \frac{\tan (45° + \tfrac{1}{2}\varphi_0)}{\tan \varphi_0} . \tag{5.103}$$

Now

$$\lim_{\varphi_0 \to 90°} \frac{\tan (45 + \tfrac{1}{2}\varphi_0)}{\tan \varphi_0} = \lim_{\varphi_0 \to 90°} \frac{1 + \tan^2 \tfrac{1}{2}\varphi_0}{1 - \tan^2 \tfrac{1}{2}\varphi_0} \times \frac{1 - \tan^2 \tfrac{1}{2}\varphi_0}{2 \tan \tfrac{1}{2}\varphi_0}$$

$$= \lim_{\varphi_0 \to 90°} \frac{(1 + \tan \tfrac{1}{2}\varphi_0)^2}{2 \tan \tfrac{1}{2}\varphi_0} = 2.$$

Thus (5.103) becomes

$$\rho = \frac{2a}{(1-\varepsilon^2)^{\frac{1}{2}}} \left(\frac{1+\varepsilon}{1-\varepsilon}\right)^{\frac{1}{2}\varepsilon} \tan{(45° - \tfrac{1}{2}\varphi)} \left(\frac{1+\varepsilon \sin \varphi}{1-\varepsilon \sin \varphi}\right)^{\frac{1}{2}\varepsilon}. \qquad (5.104)$$

The transformation formulae then become

$$\rho = 2a(1+\varepsilon)^{-\frac{1}{2}(1-\varepsilon)} (1-\varepsilon)^{-\frac{1}{2}(1+\varepsilon)} \tan{(45° - \tfrac{1}{2}\varphi)} \left(\frac{1+\varepsilon \sin \varphi}{1-\varepsilon \sin \varphi}\right)^{\frac{1}{2}\varepsilon} \quad (5.105)$$

and

$$\vartheta = \lambda.$$

Since the scale distortion $m = \rho/N \cos \varphi$, and $\partial\rho/\partial\varphi = -mM$ $\qquad (5.106)$

$$\frac{\partial m}{\partial \varphi} = \frac{mM}{N \cos \varphi}(-1 + \sin \varphi)$$

(see also (5.75) a.f.).

Thus

$$\left(\frac{\partial^2 m}{\partial \varphi^2}\right)_0 = \frac{M_0(\sin \varphi_0 - 1)}{N_0 \cos \varphi_0}\left(\frac{\partial m}{\partial \varphi}\right)_0 + \frac{m_0(\sin \varphi_0 - 1)}{N_0 \cos \varphi_0}\left(\frac{\partial M}{\partial \varphi}\right)_0 +$$

$$+ \frac{m_0(\sin \varphi_0 - 1)M_0^2 \sin \varphi_0}{N_0^2 \cos^2 \varphi_0} + m_0 \frac{M_0}{N_0}$$

or

$$\left(\frac{\partial^2 m}{\partial \varphi^2}\right)_{0\varphi_0 \to 90°} = \lim \frac{(\sin \varphi_0 - 1) \sin \varphi_0}{\cos^2 \varphi_0} + 1 = -\tfrac{1}{2} + 1 = \tfrac{1}{2}. \qquad (5.107)$$

The scale distortion therefore becomes

$$m = 1 + \tfrac{1}{4}\Delta\varphi^2. \qquad (5.108)$$

The irregularity in comparison with (5.87) originates from the apex of the cone being a singular point coinciding with the central point of the projection.

It can be derived from (5.74) or directly from (5.105) that

$$\left(\frac{\partial\rho}{\partial\varphi}\right)_0 = 2a(1+\varepsilon)^{-\frac{1}{2}(1-\varepsilon)} (1+\varepsilon)^{-\frac{1}{2}(1+\varepsilon)} \frac{1}{2}\left(\frac{1+\varepsilon}{1-\varepsilon}\right)^{\frac{1}{2}\varepsilon}$$

or

$$\left(\frac{\partial\rho}{\partial\varphi}\right)_0 = a(1+\varepsilon)^{-(\frac{1}{2}-\varepsilon)} (1-\varepsilon)^{-(\frac{1}{2}+\varepsilon)}. \qquad (5.109)$$

Formula (5.108) then becomes after transposing with (5.109)

$$m = 1 + \frac{(1+\varepsilon)^{1-2\varepsilon}(1-\varepsilon)^{1+2\varepsilon}}{4a^2} \Delta\rho^2$$

or, by (5.102)

$$m = 1 + \frac{(1+\varepsilon)^{(1-2\varepsilon)}(1-\varepsilon)^{(1+2\varepsilon)}}{4a^2}(X^2 + Y^2). \tag{5.110}$$

From (5.110) it is concluded that the curve of equal scale distortion is a circle. Similar deliberations as for the Lambert conical projections are of course valid for both these projections but they will not be pursued here.

### 5.3.4   Spherical transformation formulae

The formulae for the sphere are obtained putting $\varepsilon = 0$. They become:

(1) The Lambert conical projection, one standard parallel $\varphi_0$

$$\left.\begin{aligned} \rho &= R \cot \varphi_0 \left\{ \frac{\tan (45° - \tfrac{1}{2}\varphi)}{\tan (45° - \tfrac{1}{2}\varphi_0)} \right\}^{\sin \varphi_0} \\[2mm] \vartheta &= \lambda \sin \varphi_0 \\[2mm] m &= \frac{\rho \sin \varphi_0}{R \cos \varphi}. \end{aligned}\right\} \tag{5.111}$$

(2) The Lambert conical projection; two standard parallels $\varphi_1$, and $\varphi_2$

$$\left.\begin{aligned} \rho &= R \frac{\cos \varphi_1}{\sin \varphi_0} \left\{ \frac{\tan (45° - \tfrac{1}{2}\varphi)}{\tan (45° - \tfrac{1}{2}\varphi_1)} \right\}^{\sin \varphi_0} \\[2mm] \vartheta &= \lambda \sin \varphi_0 \\[2mm] m &= \frac{\rho \sin \varphi_0}{R \cos \varphi} \\[2mm] \sin \varphi_0 &= \frac{\ln \cos \varphi_1 - \ln \cos \varphi_2}{\ln \tan (45° - \tfrac{1}{2}\varphi_1) - \ln \tan (45° - \tfrac{1}{2}\varphi_2)}. \end{aligned}\right\} \tag{5.112}$$

In both projections rectangular coordinates are obtained by (5.67)

$$X = \rho_0 - \rho \cos \vartheta$$
$$Y = \rho \sin \vartheta .$$

(3) The Mercator projection

$$\left.\begin{aligned} X &= \ln \tan (45° + \tfrac{1}{2}\varphi) \\[4pt] Y &= R\lambda \\[8pt] m &= \frac{1}{\cos \varphi} \, . \end{aligned}\right\} \tag{5.113}$$

(4) The polar stereographic projection

$$\left.\begin{aligned} \rho &= 2R \tan (45° - \tfrac{1}{2}\varphi) \\[4pt] \vartheta &= \lambda \\[8pt] m &= \frac{\rho}{R \cos \varphi} = \frac{1}{\cos^2 (45° - \tfrac{1}{2}\varphi)} \\[8pt] X &= \rho \cos \vartheta \qquad Y = \rho \sin \vartheta \, . \end{aligned}\right\} \tag{5.114}$$

These formulae are simplified using the $(X, Y)$ coordinate system

$$\left.\begin{aligned} X &= 2R \cos \lambda \tan (45° - \tfrac{1}{2}\varphi) = 2R \frac{\cos \varphi \cos \lambda}{1 + \sin \varphi} \\[8pt] Y &= 2R \sin \lambda \tan (45° - \tfrac{1}{2}\varphi) = 2R \frac{\cos \varphi \sin \lambda}{1 + \sin \varphi} \\[8pt] m &= \frac{2}{1 + \sin \varphi} \, . \end{aligned}\right\} \tag{5.115}$$

If in restricted non-global areas the ellipsoid is depicted through the projection on the Gaussian or conformal sphere (as shown in Section 5.2) the value of $R$ in the central point is taken equal to $\sqrt{M_0 N_0}$    (5.67).

### 5.3.5   Oblique and transverse projections

Oblique and transverse conical, cylindrical and stereographic projections are given by the respective formulae (5.111) to (5.114) inclusive, if the $\lambda$ and $\varphi$ refer to the new pole $P$ and not to the North pole $NP$. In Chapter 2 these angles have been denoted by $\alpha$ and $h$ respectively. The oblique and transverse projections of the ellipsoid to the plane directly will not be discussed. Neither will the oblique and transverse conical projections be dealt with. Reference is made to the literature [2], [6], [26].

The oblique Mercator projection is suitable for the mapping of countries stretching lengthwise about a great circle. With the transverse Mercator projection this great circle is a meridian. The reference great circle (meridian) is depicted as a straight line. The stereographic projections are suitable for the mapping of circular areas.

By substitution of the formulae (2.32) and (2.33) into (5.112) the transformation formula of the oblique Mercator projection becomes

$$X = R \ln \tan (45° + \tfrac{1}{2}h) = \tfrac{1}{2} R \ln \frac{1+\sin h}{1-\sin h}$$

or

and

$$\left.\begin{aligned}
X &= \tfrac{1}{2} R \ln \frac{1+\sin \varphi \sin \varphi_P + \cos \varphi \cos \varphi_P \cos \lambda}{1-\sin \varphi \sin \varphi_P - \cos \varphi \cos \varphi_P \cos \lambda} \\[2mm]
Y &= R \arctan \frac{\sin \lambda \cos \varphi}{\cos \varphi \sin \varphi_P \cos \lambda - \cos \varphi_P \sin \varphi}.
\end{aligned}\right\} \quad (5.116)$$

The transverse Mercator projection is derived from (5.116) by putting $\varphi_P = 0$:

$$\left.\begin{aligned}
X &= \tfrac{1}{2} R \ln \frac{1+\cos \varphi \cos \lambda}{1-\cos \varphi \cos \lambda} \\[2mm]
Y &= R \arctan (-\cot \varphi \sin \lambda).
\end{aligned}\right\} \quad (5.117)$$

Analogously the oblique stereographic projection (see also (4.15) and (4.16))

$$\left.\begin{aligned}
X &= 2R \frac{\cos \alpha \cos h}{1+\sin h} = 2R \frac{\sin \varphi \cos \varphi_P - \cos \varphi \sin \varphi_P \cos \lambda}{1+\sin \varphi \sin \varphi_P + \cos \varphi \cos \varphi_P \cos \lambda} \\[2mm]
Y &= 2R \frac{\sin \alpha \cos h}{1+\sin h} = 2R \frac{\cos \varphi \sin \lambda}{1+\sin \varphi \sin \varphi_P + \cos \varphi_P \cos \varphi \cos \lambda}.
\end{aligned}\right\} \quad (5.118)$$

The mapping equations of the equatorial stereographic projection (pole $P$ on the equator; projection plane parallel to the axis of rotation) become

$$\left.\begin{aligned}
X &= 2R \frac{\sin \varphi}{1+\cos \varphi \cos \lambda} \\[2mm]
Y &= 2R \frac{\cos \varphi \sin \lambda}{1+\cos \varphi \cos \lambda}.
\end{aligned}\right\} \quad (5.119)$$

The pole $P$ in these stereographic projections coincides with the central point. It should be noted that the abscissa $X$ is still tangent to the reference great circle (meridian), directed towards the pole $P$, the $Y$ coordinate being at right angles to it.

If a new Cartesian coordinate system in the projection plane is to be adopted, $X_N$ directed to the North, the $X$, $Y$ system should be rotated as explained in Section 2.4.2.

CHAPTER 6

EQUIVALENT PROJECTIONS

## 6.1 General considerations

### 6.1.1 *Introduction*

The discussion of equivalent or equal area projections is conducted along the lines similar to those in Chapter 5 for the conformal projections though not as extensively.

It has been derived that the combination of the fundamental transformation matrix and the condition of equivalency leads to the equations (3.25)

$$
eg - f^2 = \begin{vmatrix} E' & F' \\ F' & G' \end{vmatrix} \begin{vmatrix} \dfrac{\partial U}{\partial u} & \dfrac{\partial U}{\partial v} \\ \dfrac{\partial V}{\partial u} & \dfrac{\partial V}{\partial v} \end{vmatrix}^2 . \qquad (3.25)
$$

One of the unknowns in the Jacobean determinant may be selected according to the special requirements of a particular projection, thus fixing the other unknown.

Areas remain undistorted, the scale distortion varies, and angles do not remain unchanged. As has been stated before, equivalency and conformality cannot be achieved completely at the same time. Lines and angles, however, are not entirely unimportant in the total use of an equivalent projection. Therefore their behaviour is treated in some detail. Attention is drawn to isoperimetric curves indicating the loci of points of scale distortion equal to unity. A preliminary remark may be made that the orthogonal system of meridians and parallels on the ellipsoid or the sphere is not necessarily transformed into an orthogonal system of parametric curves in the plane.

### 6.1.2    *The scale distortion: isoperimetric curves.*

The scale distortion in an arbitrary direction may be written as

$$m^2 = \frac{E \, d\varphi^2 + 2F \, d\varphi \, d\lambda + G \, d\lambda^2}{e \, d\varphi^2 + g \, d\lambda^2}, \tag{6.1}$$

taking the general case, that the orthogonal system $(\varphi, \lambda)$ on the ellipsoid is not transformed as an orthogonal system in the plane.

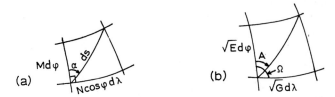

Fig. 6.1
(a) Datum plane, (b) image plane

Referring to Section 3.2.1 and Figure 6.1 it is seen that (6.1) becomes

$$m^2 = \frac{E \, d\varphi^2 + 2\sqrt{EG} \cos \Omega \, d\varphi \, d\lambda + G \, d\lambda^2}{M^2 \, d\varphi^2 + N^2 \cos^2 \varphi \, d\lambda^2} \tag{6.2}$$

with

$$\cos \Omega = \frac{F}{\sqrt{EG}} \tag{6.3}$$

$$\tan \alpha = \frac{N \cos \varphi}{M} \frac{d\lambda}{d\varphi} \tag{6.4}$$

and

$$\tan A = \frac{\sqrt{G} \sin \Omega \, d\lambda}{\sqrt{E} \, d\varphi + \sqrt{G} \cos \Omega \, d\lambda}. \tag{6.5}$$

Substituting (6.4) into (6.5):

$$\tan A = \frac{M\sqrt{G} \tan \alpha \sin \Omega}{N \cos \varphi \sqrt{E} + M\sqrt{G} \tan \alpha \cos \Omega} \tag{6.6}$$

The projections concerned being equivalent

$$MN \cos \varphi \, d\varphi \, d\lambda = \sqrt{EG} \sin \Omega \, d\varphi \, d\lambda$$

or

$$\sin \Omega = \frac{MN \cos \varphi}{\sqrt{EG}} = \frac{1}{m_\varphi m_\lambda}. \qquad (6.7)$$

Hence, after some deductions

$$\tan A = \frac{M^2 \tan \alpha}{E + M^2 \tan \alpha \cot \Omega} = \frac{\tan \alpha}{m_\varphi^2 + \tan \alpha \cot \Omega}. \qquad (6.8)$$

This is the general relationship between the azimuth of an arc on the ellipsoid and in the projection. An explicit function of $(A - \alpha)$ becomes rather complicated.

Tissot's indicatrix dealt with in Section 3.5 indicates the scale distortion in an arbitrary direction. The scale distortion in the direction of the projected meridians and parallels $m_\varphi$ and $m_\lambda$ respectively are conjugate diameters of the ellipse, the relationships having been deduced in Section 3.5.2. The main ones are, in the present notation and adapted for equivalent projections:

$$m_\varphi^2 = m_0^2 \cos^2 \alpha_\varphi + m_{90}^2 \sin^2 \alpha_\varphi$$

$$m_\lambda^2 = m_0^2 \sin^2 \alpha_\varphi + m_{90}^2 \cos^2 \alpha_\varphi$$

whence

$$m_\lambda^2 + m_\varphi^2 = m_0^2 + m_{90}^2$$

$$m_\lambda m_\varphi \sin \Omega = m_0 m_{90} = 1.$$

In an arbitrary direction, counted from the $m_0$ axis

$$m^2 = m_0^2 \cos^2 \alpha + m_{90}^2 \sin^2 \alpha.$$

The scale distortion in one particular direction becomes equal to unity if

$$m_0^2 \cos^2 \alpha + m_{90}^2 \sin^2 \alpha = 1 = m_0 m_{90}$$

so that

$$m_{90}^2 \sin^2 \alpha = m_0^2 \cos^2 \alpha$$

or

$$\tan \alpha = \pm \sqrt{\frac{m_0}{m_{90}}}. \qquad (6.9)$$

In the case of equivalent projections the possibility that $\sqrt{m_0/m_{90}} = 1$ with $m_0 m_{90} = 1$ can only arise if

$$\tan \alpha = \pm \sqrt{m_0^2} = \pm m_0 \qquad (6.10)$$

Now recall the formulae for the length distortion $m_{\bar{3}}$ of the bearing of the maximum angular distortion $\zeta$, Section 3.4.5.

Comparing these formulae with a consideration of (6.10) it is clear that

$$\alpha = \pm (45° - \tfrac{1}{2}\zeta) . \tag{6.11}$$

The corresponding angle $A$ in the projection is then equal to

$$A = \pm (45° + \tfrac{1}{2}\zeta) .$$

If at every point the major and minor axes of the indicatrix are drawn, and the bearings $\alpha$ and $-\alpha$ (or $A$ and $-A$) of (6.11) are set off, the locus of points with scale distortion $m = 1$ will be a curve running approximately in the directions of the axes system of the indicatrix. These curves are called *isoperimetric* curves.

The differential equations of the isoperimetric curves for the individual projections dealt with in the following sections will not be derived. Reference is made to the specific literature [1], [15].

## 6.2   The equivalent projection of the ellipsoid on the sphere

### 6.2.1   *Introduction*

For the same reasons as with the conformal projections, it is an advantage to transform the ellipsoid on to a sphere of the same area. In this projection the longitude on both surfaces is preserved. The latitude on the sphere is called the "authalic" latitude, a name invented by Tissot [27] and frequently applied in American bibliography on the subject. Driencourt [7] uses "equivalent latitude".

The radius of the sphere is called correspondingly the authalic or equivalent radius $(R_A)$.

The relationships between the ellipsoidal and spherical elements are

$$\lambda = \Lambda$$

$$\left.\begin{aligned} R_A \sin \Phi_A &= \tfrac{1}{2}b^2 \left\{ \frac{\sin \varphi}{1 - \varepsilon^2 \sin^2 \varphi} + \frac{1}{2\varepsilon} \ln \frac{1 + \varepsilon \sin \varphi}{1 - \varepsilon \sin \varphi} \right\} \\[2ex] R_A^2 &= \tfrac{1}{2}b^2 \left\{ \frac{1}{1 - \varepsilon^2} + \frac{1}{2\varepsilon} \ln \frac{1 + \varepsilon}{1 - \varepsilon} \right\} \end{aligned}\right\} \tag{6.12}$$

where $b$ is the semi minor axis of the meridional ellipse. These expressions may be expanded into power series.

### 6.2.2   *Transformation formulae. Authalic latitude and radius*

The transformation formulae for this projection are deduced from the line elements on both surfaces; the condition of equivalency have been developed in (3.25) and restated in Section 6.1.1.

On the ellipsoid:

$$ds^2 = M^2 \, d\varphi^2 + N^2 \cos^2 \varphi \, d\lambda^2 , \tag{6.13}$$

whence $eg - f^2 = M^2 N^2 \cos^2 \varphi$ .

On the sphere

$$dS^2 = R_A^2 \, d\Phi_A^2 + R_A^2 \cos^2 \Phi_A \, d\Lambda$$

with

$$E' \, G' - F' = R_A^4 \cos^2 \Phi_A . \tag{6.14}$$

Inserting (6.13) and (6.14) into (3.25) gives

$$M^2 N^2 \cos^2 \varphi = R_A^4 \cos^2 \Phi_A \begin{vmatrix} \dfrac{\partial \Phi_A}{\partial \varphi} & \dfrac{\partial \Phi_A}{\partial \lambda} \\[2mm] \dfrac{\partial \Lambda}{\partial \varphi} & \dfrac{\partial \Lambda}{\partial \lambda} \end{vmatrix}^2 . \tag{6.15}$$

Since it was imposed that

$$\lambda = \Lambda \tag{6.16}$$

and $\Phi_A$ should be independent of $\Lambda$ as is the case with $\varphi$ and $\lambda$, consequently

$$\frac{\partial \Lambda}{\partial \lambda} = 1; \quad \frac{\partial \lambda}{\partial \varphi} = \frac{\partial \Phi_A}{\partial \lambda} = 0 .$$

Hence (6.15) becomes

$$M^2 N^2 \cos^2 \varphi = R_A^4 \cos^2 \Phi_A \begin{vmatrix} \dfrac{\partial \Phi_A}{\partial \varphi} & 0 \\[2mm] 0 & 1 \end{vmatrix}^2$$

or

$$MN \cos \varphi = R_A^2 \cos \Phi_A \frac{\partial \Phi_A}{\partial \varphi} . \tag{6.17}$$

Working this out, it is found that

$$R_A^2 \sin \Phi_A = a(1 - \varepsilon^2) \int \frac{\cos \varphi}{(1 - \varepsilon^2 \sin^2 \varphi)} \, d\varphi + c. \tag{6.18}$$

This leads to the solution

$$R_A^2 \sin \Phi_A = b^2 \left( \frac{\sin \varphi}{2(1 - \varepsilon^2 \sin^2 \varphi)} + \frac{1}{4\varepsilon} \ln \frac{1 + \varepsilon \sin}{1 - \varepsilon \sin} \right) + c. \tag{6.19}$$

Since for $\varphi = 0$, $\Phi_A = 0$, the constant is also equal to zero. Frequently this formula is expanded in powers of $\varepsilon$ and $\sin \varphi$:

$$\frac{\sin \varphi}{2(1 - \varepsilon^2 \sin^2)} = \tfrac{1}{2}(\sin \varphi + \varepsilon^2 \sin^3 \varphi + \varepsilon^4 \sin^5 \varphi + \ldots \tag{6.20}$$

and

$$\left. \begin{aligned} \frac{1}{4\varepsilon} \ln \frac{1 + \varepsilon \sin \varphi}{1 - \varepsilon \sin \varphi} &= \frac{1}{4\varepsilon} \{\ln (1 + \varepsilon \sin \varphi) - \ln (1 - \varepsilon \sin \varphi)\} = \\ &= \tfrac{1}{2}(\sin \varphi + \tfrac{1}{3}\varepsilon^2 \sin^3 \varphi + \tfrac{1}{5}\varepsilon^4 \sin^5 \varphi + \ldots \end{aligned} \right\} \tag{6.21}$$

By substitution of (6.20) and (6.21) into (6.19) we get

$$R_A^2 \sin \Phi_A = b^2 \sin \varphi (1 + \tfrac{2}{3}\varepsilon^2 \sin^2 \varphi + \tfrac{3}{5}\varepsilon^4 \sin^4 \varphi + \tfrac{4}{7}\varepsilon^6 \sin^6 \varphi + \ldots \tag{6.22}$$

At the pole $\varphi = \Phi_A = 90°$ which fact renders $R_A$ from (6.19)

$$R_A^2 = b^2 \left( \frac{1}{2(1 - \varepsilon^2)} + \frac{1}{4\varepsilon} \ln \frac{1 + \varepsilon}{1 - \varepsilon} \right) \tag{6.23}$$

or in an expanded form

$$R_A^2 = b^2 (1 + \tfrac{2}{3}\varepsilon^2 + \tfrac{3}{5}\varepsilon^4 + \tfrac{4}{7}\varepsilon^6 + \ldots) \tag{6.24}$$

By dividing (6.19) by (6.23) or (6.22) by (6.24) an expression for $\Phi_A$ is obtained. In the latter form of an expansion

$$\left. \begin{aligned} \sin \Phi_A &= \sin \varphi \frac{1 + \tfrac{2}{3}\varepsilon^2 \sin^2 \varphi + \tfrac{3}{5}\varepsilon^2 \sin^4 \varphi + \ldots}{1 + \tfrac{2}{3}\varepsilon^2 + \tfrac{3}{5}\varepsilon^4 + \tfrac{4}{7}\varepsilon^6 + \ldots} \\ &= 1 - \tfrac{2}{3}\varepsilon^2 \cos^2 \varphi + \ldots \end{aligned} \right\} \tag{6.25}$$

## 6.3    Equivalent projections on the cone, the cylinder and the plane

### 6.3.1    *Introduction*

There are numerous equivalent projections, many of which are described in detail in [1] and [15].

A normal conical, a cylindrical and an azimuthal type of projection will be discussed as in Chapter 5 for conformal projections.

Since the interest is largely directed towards the equivalent image of large areas, the refinement of projecting the ellipsoid is less important than with the conformal projections. Of course one can always introduce the equivalent latitude and radius of the authalic sphere.

Most of the equivalent projections try to minimize the length distortion, or to balance length and angle distortions.

Lambert has devised an equivalent conical projection with one standard parallel. So has Albers. In both projections the meridians are converging straight lines, the parallels are concentric circles with the centre at the point of convergence. The scale-distortion on the standard parallel is equal to unity. There are, however, significant differences in the distribution of length and angle distortions.

Albers also devised a conical projection with two standard parallels on which a true scale is maintained.

The normal cylindrical equivalent projection (also by Lambert) has equally spaced parallel meridians with parallel straight lines of equal latitude spaced proportional to $\sin \varphi$. There are considerable angular distortions.

The Lambert (polar) equivalent azimuthal projection is derived directly from the conical projection. The meridians are straight lines converging at the pole; the parallels are concentric circles about it. The projection is particularly useful for atlases.

Formulae for the corresponding transverse and oblique projections are not given here. If necessary they can be readily derived by a rotation of the coordinate system on the sphere.

### 6.3.2   Conical equivalent projections (Lambert and Albers)

The elements necessary for the formulae of an equivalent conical transformation are displayed in Figure 6.2.

The line element on the sphere is indicated as usual by

$$ds^2 = R^2 \, d\varphi^2 + R^2 \cos^2\varphi \, d\lambda \tag{6.26}$$

with

$$eg = R^4 \cos^2 \varphi$$

where $R$ and $\varphi$ are meant to be, if necessary, the authalic radius and latitude respectively, the suffix $A$ having been omitted for convenience.

In the plane a polar coordinate system is adopted

$$dS^2 = d\rho^2 + \rho^2 d \vartheta^2$$

with

$$E' G' = \rho^2 \, .$$

The central point of the projection is $O$ with coordinates $(\varphi_0 \, \lambda_0)$ (not necessarily on the tangent circle). The reference meridian is $AO$.

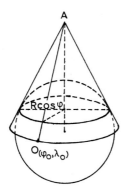 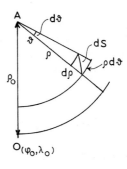

Fig. 6.2
Equivalent conical projection

The first condition is that $\vartheta$ is independent of $\varphi$, the second one that $\rho$ is independent of $\lambda$.
Thus

$$\rho = f(\varphi) \tag{6.27}$$
$$\vartheta = c_1 \lambda + c_2 \, .$$

with $\partial \vartheta / \partial \lambda = c_1$

The constant $c_2$ is equal to zero, if for $\lambda = 0$, $\vartheta = 0$. The condition of equivalency then becomes

$$R^2 \cos \varphi = \rho \begin{vmatrix} \dfrac{\partial \rho}{\partial \varphi} & 0 \\[2ex] 0 & c_1 \end{vmatrix} \tag{6.28}$$

or

$$R^2 \cos \varphi = -\rho_1 c \frac{\partial \rho}{\partial \varphi}$$

(an increment of $\varphi$ corresponds to a decrease of $\rho$), whence

$$\rho^2 = -\frac{2 R^2 \sin \varphi}{c_1} + c_3 \, . \tag{6.29}$$

Equations (6.27) and (6.29) are the basic formulae of the conical equivalent projections.

The constant $c_1$ is a scale factor and $c_3$ depends on the boundary conditions to be imposed.

The scale distortion

$$m_{90}^2 = \frac{c_1^2 \rho^2}{R^2 \cos^2 \varphi} \tag{6.30}$$

(see (3.90), or by noting that $\rho \, d\vartheta = \rho m_{90} \, d\lambda$).

or

$$m_{90} = \frac{c_1 \rho}{R \cos \varphi} \; ; \tag{6.31}$$

and because of the equivalency

$$m_0 = \frac{R \cos \varphi}{c_1 \rho} . \tag{6.32}$$

The scale distortion in a particular azimuth $\vartheta$ is obtained by formula (3.47). Plane Cartesian coordinates are calculated by

$$X = \rho_0 - \rho \cos \vartheta$$

$$Y = \rho \sin \vartheta .$$

The constant $c_1$ gives additional freedom in imposing special conditions on the scale distortion for further refined features of the map, e.g. preservation of the scale and a zero angular distortion along the parallel on which the central point is selected, or a smallest angular distortion averaged out on the region etc.

Albers selected the case in which the central point is located on the parallel, where the cone is tangent to the sphere. Then

$$\left.\begin{array}{l} \rho_0 = R \cot \varphi_0 \\[2mm] \vartheta = \dfrac{R \cos \varphi_0}{R \cot \varphi_0} \lambda = \lambda \sin \varphi_0 \end{array}\right\} \tag{6.33}$$

so

$$\frac{\partial \vartheta}{\partial \lambda} = \sin \varphi_0 = c_1 , \tag{6.34}$$

equal to the constant of the cone.

Substitution of (6.34) into (6.29) yields

$$\rho^2 = -\frac{2R^2 \sin \varphi}{\sin \varphi_0} + c_3 . \tag{6.35}$$

In the central point

$$\rho_0 = -2R^2 + c_3$$

whence

$$c_3 = R^2(2 + \cot^2 \varphi_0) . \tag{6.36}$$

Solving for $\rho$ gives

$$\rho^2 = R^2\left(1 + \frac{1}{\sin^2 \varphi_0} - \frac{2 \sin \varphi}{\sin \varphi_0}\right)$$

or

$$\rho = \frac{R}{\sin \varphi_0}\sqrt{1 + \sin^2 \varphi_0 + 2 \sin \varphi \sin \varphi_0} . \tag{6.37}$$

The transformation formulae therefore are given by (6.37) and

$$\vartheta = \lambda \sin \varphi_0 . \tag{6.38}$$

A remarkable feature of this projection is that the pole ($\varphi = 90°$) is transformed into a circle the radius of which is equal to

$$\rho_{\text{Pole}} = \frac{R}{\sin \varphi_0}(1 - \sin \varphi_0) = \frac{R}{\sin \varphi_0}\sin (45° - \tfrac{1}{2}\varphi_0) , \tag{6.39}$$

and the cone thus is truncated.

The scale distortions along the meridians and parallels are

$$\left.\begin{aligned} m_0 &= \frac{R \cos \varphi}{\rho \sin \varphi_0} \\[2mm] m_{90} &= \frac{\rho \sin \varphi_0}{R \cos \varphi} \end{aligned}\right\} . \tag{6.40}$$

and

On the standard parallel $m_0 = m_{90} = 1$ (conformality). This was the Albers equivalent conical projection with one standard parallel. As with the Lambert conformal conical projection, two standard parallels may be selected with scale distortion equal to unity.

The radii of these parallels are (see (6.31))

$$\left.\begin{aligned} \rho_1 c_1 &= R \cos \varphi_1 \\ \rho_2 c_1 &= R \cos \varphi_2 . \end{aligned}\right\} . \tag{6.41}$$

Also

$$\left.\begin{array}{l} \vartheta = c_1 \lambda = \dfrac{R \cos \varphi_1}{\rho_1} \lambda \\[12pt] \vartheta = c_1 \lambda = \dfrac{R \cos \varphi_2}{\rho_2} \end{array}\right\}. \qquad (6.42)$$

Solving $\rho_1$ and $\rho_2$ from (6.41) and substituting the result into (6.29) gives

$$\frac{R^2 \cos^2 \varphi_1}{c_1^2} = -\frac{2 R^2 \sin \varphi_1}{c_1} + c_3$$

and

$$\frac{R^2 \cos^2 \varphi_2}{c_1^2} = -\frac{2 R^2 \sin \varphi_2}{c_1} + c_3$$

or

$$R^2 \cos^2 \varphi_1 + 2 R^2 c_1 \sin \varphi_1 - c_1^2 c_3 = 0$$

$$R^2 \cos^2 \varphi_2 + 2 R^2 c_1 \sin \varphi_2 + c_1^2 c_3 = 0 .$$

The constant $c_1$ may be selected from these last equations by subtraction and division

$$c_1 = \frac{R^2 (\cos^2 \varphi_1 - \cos^2 \varphi_2)}{2 R^2 (\sin \varphi_2 - \sin \varphi_1)} = \tfrac{1}{2}(\sin \varphi_1 + \sin \varphi_2) . \qquad (6.43)$$

Hence the transformation formulae become expressing $\rho$ in terms of $\rho_1$ and $\rho_2$:

$$\rho^2 = \rho_1^2 + \frac{4 R^2 (\sin \varphi_1 - \sin \varphi)}{(\sin \varphi_1 + \sin \varphi_2)} \qquad (6.44)$$

with a check

$$\rho^2 = \rho_2^2 + \frac{4 R^2 (\sin \varphi_2 - \sin \varphi)}{(\sin \varphi_1 + \sin \varphi_2)}$$

and

$$\vartheta = \tfrac{1}{2}(\sin \varphi_1 + \sin \varphi_2) \lambda . \qquad (6.45)$$

The scale distortions $m_0$ and $m_{90}$ are

$$\left.\begin{array}{l} m_0 = \dfrac{2 R \cos \varphi}{\rho (\sin \varphi_1 + \sin \varphi_2)} \\[14pt] m_{90} = \dfrac{\rho (\sin \varphi_1 + \sin \varphi_2)}{2 R \cos \varphi} . \end{array}\right\} \qquad (6.46)$$

When speaking of the Albers projection, usually the latter one with two standard parallels is referred to, because of its greater importance in practical applications.

### 6.3.3  Lambert's cylindrical and azimuthal equivalent projections

Lambert's normal equal cylindrical projection is conveniently found by taking a plane rectangular coordinate system

$$dS^2 = dX^2 + dY^2$$

with

$$E' G' = 1 .$$

(6.47)

The meridians being taken parallel to the $X$-axis and straight lines, the relationship

$$Y = f(\lambda) = R\lambda$$

is selected. Using formulae (6.26) the condition of equivalency is obtained by

$$R^4 \cos^2 \varphi = \begin{vmatrix} \dfrac{\partial X}{\partial \varphi} & \cdots \\ 0 & R \end{vmatrix}^2 = R^2 \left( \dfrac{\partial X}{\partial \varphi} \right)^2$$

(6.48)

or

$$R \cos \varphi = \dfrac{\partial X}{\partial \varphi}$$

(6.49)

By integration

$$X = R \sin \varphi + c .$$

The constant $c$, however, is equal to zero by taking the central point on the equator ($X = 0$ for $\varphi = 0$).
The projection formulae are

$$\begin{aligned} X &= R \sin \varphi \\ Y &= R\lambda \end{aligned}$$

(6.50)

Further

$$\begin{aligned} m_0 &= \cos \varphi \\ m_{90} &= \dfrac{1}{\cos \varphi} \end{aligned}$$

(6.51)

and the maximum angular distortion (3.57)

$$\sin \zeta = \frac{m_{90} - m_0}{m_{90} + m_0} = \frac{1 - \cos^2 \varphi}{1 + \cos^2 \varphi} .$$    (6.52)

This projection is conformal and equidistant on the equator. An insight on the construction may be given by Figure 6.3a and b.

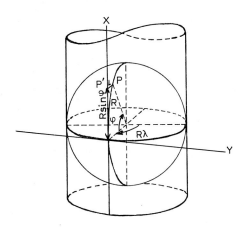

Fig. 6.3a
Geometrical construction of Lambert's equivalent cylindrical projection

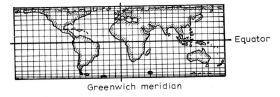

Fig. 6.3b
Lambert's equivalent cylindrical projection

The Lambert azimuthal equivalent projection (Fig. 6.4) is derived directly from the conical projection by putting $\sin \varphi_0 = \sin 90° = 1$ into (6.37) and (6.38), the transformation formulae becoming

$$\left. \begin{array}{l} \rho = 2R \sin (45° - \tfrac{1}{2} \varphi) \\ \vartheta = \lambda . \end{array} \right\}$$    (6.53)

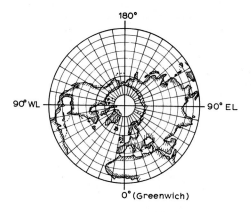

Fig. 6.4
Lambert's equivalent azimuthal projection

Also

$$m_0 = \frac{R \cos \varphi}{\rho} = \cos (45° - \tfrac{1}{2}\varphi) \Bigg\}$$

$$m_{90} = \frac{\rho}{R \cos \varphi} = \frac{1}{\cos (45° - \tfrac{1}{2}\varphi)} \Bigg\}$$

(6.54)

and

$$\sin \zeta = \frac{1 - \cos^2 (45° - \tfrac{1}{2}\varphi)}{1 + \cos^2 (45° - \tfrac{1}{2}\varphi)}.$$

(6.55)

There is no angular distortion in the central point ($\varphi_0 = 90°$) Rectangular coordinates are obtained by (5.102)

$$X = \rho \cos \vartheta \Bigg\}$$
$$Y = \rho \sin \vartheta .$$

(5.102)

## 6.4   Bonne's pseudo conical equivalent projection; the Sanson–Flamsteed and Werner projections

### 6.4.1   Introduction

At the conclusion of this chapter Bonne's equivalent projection is treated as an example of a pseudo or modified conical projection. Only the reference central meridian is a straight line, being a section of the cone and the central meridional plane. The meridians and parallel circles do not form an orthogonal system in the image plane. The scale distortion on the central meridian and along all parallels is equal to unity. It is not preserved along the other meridians, The distances $S$ between the parallels $\varphi_1$ and $\varphi_2$ (which are projected as circles concentric about the apex of the cone) is

$$S = \int_{\varphi_1}^{\varphi_2} M \, d\varphi. \qquad (6.56)$$

The parallel circles are therefore not equally spaced, $M$ being different at various latitudes. However, in the following derivations spherical expressions

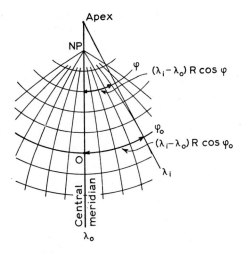

Fig. 6.5a

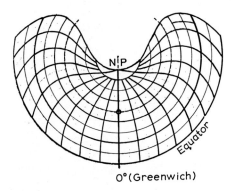

0°(Greenwich)

Fig. 6.5b
Hemisphere in Bonne's projection ($\varphi_0 = 45°$)

will be used. The distance set off along a parallel at latitude $\varphi$ from the central meridian $\lambda_0$ up to the meridian $\lambda_i$ is equal to

$$(\lambda_i - \lambda_0)N \cos \varphi . \tag{6.57}$$

This is illustrated in Figure 6.5, where the parametric network is drawn in Bonne's projection.

The principal axes of the Tissot indicatrix do not coincide with the $(\varphi, \lambda)$ system of parametric lines. This necessitates additional calculations in order to find the $m_0$ and $m_{90}$ in the manner as shown in Sections 3.5.2 and 6.1.2.

The significance of the Bonne projection for medium scale maps is greatly reduced at the present. The distortions in angles and distances are such that the projection cannot compete successfully with other projections having better balanced features from the point of view of geodetic and topographical applications.

In the manner already known, the apex of the cone can be moved into infinity along the rotational axis. One then obtains a pseudo cylindrical equivalent projection named after Sanson and Flamsteed. The parallel circles become straight parallel lines ($\rho = \infty$), equally spaced according to (6.56). By setting off distances according to (6.57) the meridians become sinusoidal curves. The formulae and an illustration are given in Section 6.4.3. In the same section, a polar equivalent projection is described where the central point coincides with the pole. Maintaining the same properties as Bonne's the parallel circles remain circular, concentric about the pole. Then $\rho_0 = R \cot \varphi_0 = 0$, the radius $\rho$ becoming zero for $\varphi_0 = 90°$. This is Werner's equivalent projection. Both the Sanson – Flamsteed and Werner's are derived directly from Bonne's.

### 6.4.2   Bonne's projection. Indicatrix

In Bonne's projection the radius of the parallel circle $\varphi$ is equal to

$$\rho = \rho_0 - \int_{\varphi_0}^{\varphi} R \, d\varphi \tag{6.58}$$

with

$$\left. \begin{aligned} \frac{\partial \rho}{\partial \varphi} &= -R \\[2em] \frac{\partial \rho}{\partial \lambda} &= 0 \end{aligned} \right\} \tag{6.59}$$

and

The condition of equivalency then becomes

$$R^4 \cos^2 \varphi = \rho^2 \begin{vmatrix} -R \dfrac{\partial \vartheta}{\partial \varphi} & \left(\dfrac{\partial \vartheta}{\partial \varphi}\right)^2 \\[2ex] 0 & \dfrac{\partial \vartheta}{\partial \lambda} \end{vmatrix} = R^2 \rho^2 \left(\dfrac{\partial \vartheta}{\partial \lambda}\right)^2 \tag{6.60}$$

or

$$R \cos \varphi = \rho \left(\dfrac{\partial \vartheta}{\partial \lambda}\right).$$

Hence by integration

$$\vartheta = \frac{\lambda R \cos \varphi}{\rho} + c. \tag{6.61}$$

Assuming $\lambda = 0$ or $\vartheta = 0$ the constant $c$ is equal to zero.

The formula (6.61) signifies, what has been stated in the introduction, that on every parallel circle a distance $\lambda_i R \cos \varphi$ is to be set off from the central meridian in order to find the projection of the meridian $\lambda_i$. Consequently this meridian becomes a curve which does not intersect any parallel circles at right angles but the standard parallel $\varphi_0$. (Only the straight central meridian is orthogonal to all parallel circles.) This is readily corroborated by the equation for $F$ in the fundamental transformation matrix (3.21)

$$F = \frac{\partial \rho}{\partial \varphi} \frac{\partial \rho}{\partial \lambda} + \rho^2 \frac{\partial \vartheta}{\partial \varphi} \frac{\partial \vartheta}{\partial \lambda}. \tag{6.62}$$

Now

$$\frac{\partial \vartheta}{\partial \varphi} = -\frac{\lambda R}{\rho}\left(\sin \varphi - \frac{R \cos \varphi}{\rho}\right)$$

and

$$\frac{\partial \vartheta}{\partial \lambda} = \frac{R \cos \varphi}{\rho}. \tag{6.63}$$

Recollecting (6.59)

$$F = -\lambda R^2 \cos \varphi \left(\sin \varphi - \frac{R \cos \varphi}{\rho}\right). \tag{6.64}$$

This fundamental value becomes equal to zero

(1) on the central meridian, where $\lambda = 0$

(2) If $\varphi = 90°$

(3) If $\sin \varphi - (R \cos \varphi / \rho) = 0$, or $\rho = \rho_0 = R \cot \varphi_0$ .

This is the standard parallel. At the same time the cone is tangent to the sphere along that parallel.

In a similar manner one finds

$$E = R^2 \left\{ 1 + \lambda^2 \left( \sin \varphi - \frac{R \cos \varphi}{\rho} \right)^2 \right\}$$

$$G = R^2 \cos^2 \varphi.$$

(6.65)

It follows immediately that

$$m_\varphi = \sqrt{\frac{E}{e}} = \sqrt{1 + \lambda^2 \left( \sin \varphi - \frac{R \cos \varphi}{\rho} \right)^2}$$

(6.66)

and

$$m_\lambda = \sqrt{\frac{G}{g}} = 1 .$$

(6.67)

Remembering that

$$m_\varphi m_\lambda \sin \Omega = 1 ,$$

$$m_\varphi = \frac{1}{\sin \Omega} ,$$

(6.68)

where $\Omega$ is the angle between the meridian and the parallel counted from the meridian eastward (see Fig. 6.6).

The semi major and minor axes of the indicatrix are found by the application of (3.82)

$$(m_{90} + m_0)^2 = 4 + \cot^2 \Omega$$

$$(m_{90} - m_0)^2 = \cot^2 \Omega$$

(6.69)

whence

$$m_0 = -\tfrac{1}{2} \cot \Omega + \tfrac{1}{2} \sqrt{4 + \cot^2 \Omega}$$

$$m_{90} = \tfrac{1}{2} \cot \Omega + \tfrac{1}{2} \sqrt{4 + \cot^2 \Omega} .$$

(6.70)

The orientation of these axes is computed by (3.83) or by finding the maximum and minimum values of $\vartheta$ in the equation (6.2) after being transposed into

$$m^2 = \frac{m_\varphi^2 + 2 m_\lambda m_\varphi \cos \Omega \tan \vartheta + m_\lambda^2}{1 + \tan^2 \vartheta}$$

or

$$m^2 = \frac{m_\varphi^2 + 2 m_\varphi \cos \Omega \tan \vartheta + 1}{1 + \tan^2 \vartheta}. \qquad (6.71)$$

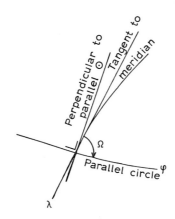

Fig. 6.6
Angle between meridian and parallel circle

It is then seen that (using (6.70))

$$\left.\begin{array}{l} \tan \vartheta_1 = -\tfrac{1}{2} \cot \Omega + \tfrac{1}{2}\sqrt{4 + \cot^2 \Omega} = m_0 \\[2mm] \tan \vartheta_2 = -\tfrac{1}{2} \cot \Omega - \tfrac{1}{2}\sqrt{4 + \cot^2 \Omega} = -m_{90} \end{array}\right\} \qquad (6.72)$$

with

and

$$\left.\begin{array}{l} \tan \vartheta_1 \tan \vartheta_2 = -1 \\[2mm] \vartheta_2 - \vartheta_1 = 90° . \end{array}\right\} \qquad (6.73)$$

These bearings have the property of being the bearings of the maximum distortion $\bar{\theta} - \vartheta = \zeta$.

Refer to formulae (3.54), (3.56) and (3.57) for the proof.

Now

$$\frac{\tan \vartheta_2 + \tan \vartheta_1}{\tan \vartheta_2 - \tan \vartheta_1} = \frac{\sin (\vartheta_2 + \vartheta_1)}{\sin (\vartheta_2 - \vartheta_1)} = \frac{-m_{90} + m_0}{-m_{90} - m_0} = \left.\begin{matrix} \\ \\ \\ \\ \\ \end{matrix}\right\}$$

$$= - \sin \zeta = - \frac{\sin \overline{\Theta} - \vartheta)}{\sin (\overline{\Theta} + \vartheta)} . \qquad (6.74)$$

Since $\vartheta_2 - \vartheta_1 = 90° = \overline{\theta} + \overline{\vartheta}$, this equation is satisfied if $\vartheta_2 + \vartheta_1 = - \overline{\theta} + \overline{\vartheta}$. Solving for $\vartheta_1$ and $\vartheta_2$ yields

$$\left.\begin{aligned} \vartheta_1 &= -45° - \tfrac{1}{2}\zeta \\ \vartheta_2 &= 45° - \tfrac{1}{2}\zeta . \end{aligned}\right\} \qquad (6.75)$$

The corresponding angles $\Theta_1$ and $\Theta_2$ in the projection are calculated with the aid of formula (6.8),

$$\tan \Theta = \frac{\tan \vartheta}{m_\varphi^2 + \tan \vartheta \cot \Omega}. \qquad (6.76)$$

Substitution of (6.68) and the second equation of (6.69) into (6.76) gives

$$\tan \Theta = \frac{\tan \vartheta}{(m_{90} - m_0) \tan \vartheta + m_{90}^2 + m_0^2 - 1} . \qquad (6.77)$$

Hence

$$\tan \Theta_1 = \frac{m_0}{m_{90}^2} \left.\begin{matrix} \\ \\ \\ \\ \\ \end{matrix}\right\}$$

and

$$\tan \Theta_2 = - \frac{m_{90}}{m_0^2} \left.\begin{matrix} \\ \\ \\ \\ \end{matrix}\right\} \qquad (6.78)$$

with $\tan \Theta_1 \tan \Theta_2 = -1$.

The angles counted from the parallel circle become somewhat simpler. Then, after some simple trigonometric transformations

$$\left.\begin{aligned} \tan (\Omega - \Theta_1) &= m_0 = - \tan (45° + \tfrac{1}{2}\zeta) \\ \tan (\Omega - \Theta_2) &= - m_{90} = \tan (45° - \tfrac{1}{2}\zeta) \end{aligned}\right\} \qquad (6.79)$$

The position of an indicatrix west of the central meridian is illustrated in the figure 6.7. The indicatrix east of this meridian is found by the formulae above, taking the sign of the functions of $\Omega > 90°$ into account.

The semi major and semi minor axes may also be expressed in terms of the maximum angular distortion.

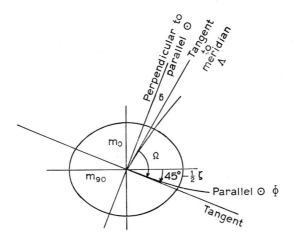

Fig. 6.7
Indicatrix

According to (3.57)

$$\sin \zeta = \frac{m_{90} - m_0}{m_{90} + m_0}$$

whence, with (6.69)

$$\tan \zeta = \frac{m_{90} - m_0}{2\sqrt{m_0 m_{90}}} = \tfrac{1}{2} \cot \Omega. \qquad (6.80)$$

By substitution of (6.80) into (6.70) it is seen that

$$\left. \begin{aligned} m_0 &= \frac{1 - \sin \zeta}{\cos \zeta} \\[2mm] m_{90} &= \frac{1 + \sin \zeta}{\cos \zeta}. \end{aligned} \right\} \qquad (6.81)$$

Finally a convenient formula is derived for the angle $\delta = 90 - \Omega$ (see Fig. 6.7), formed by the perpendicular to the parallel circle, and the tangent to the meridian at a point $P$.

The differential formula for $\delta$ is

$$\tan \delta = \frac{\rho \, d\vartheta}{R \, d\varphi}. \qquad (6.82)$$

Since

$$\frac{d\vartheta}{d\varphi} = -\frac{\lambda R}{\rho} \left( \sin \varphi - \frac{R \cos \varphi}{\rho} \right),$$

(See also 6.63)

$$\tan \delta = -\lambda \sin \varphi + \frac{R\lambda \cos \varphi}{\rho}.$$

This becomes with (2.32) and (6.61)

$$\tan \delta = \vartheta - \gamma. \tag{6.83}$$

where $\gamma$ is the convergence of meridians.

Also, with (6.80)

$$\tan \zeta = \tfrac{1}{2} \cot \Omega = \tfrac{1}{2} \tan \delta$$

and, $\zeta$ and $\delta$ being small angles, by approximation

$$\zeta = \tfrac{1}{2} \delta. \tag{6.84}$$

### 6.4.3  Sanson–Flamsteed and Werner's projections

The Sanson-Flamsteed projection is derived from Bonne's by putting $\varphi_0 = 0$. Then $\rho \to \infty$. In analogy with a previous procedure, the formulae are derived directly by taking

$$dS^2 = dX^2 + dY^2.$$

in the image plane.

Along the meridian

$$X = R\varphi$$

with

$$\frac{\partial X}{\partial \varphi} = R \quad \text{and} \quad \frac{\partial X}{\partial \lambda} = 0. \tag{6.85}$$

Then

$$R^4 \cos^2 \varphi = \begin{vmatrix} R & \dfrac{\partial Y}{\partial \varphi} \\ 0 & \dfrac{\partial Y}{\partial \lambda} \end{vmatrix}^2$$

or

$$R \cos \varphi = \frac{\partial Y}{\partial \lambda}. \tag{6.86}$$

By integration

$$Y = \lambda R \cos \varphi + c \qquad (6.87)$$

where $c = 0$   for   $Y = 0$   if   $\lambda = 0$.
   From (6.87)

$$\frac{\partial Y}{\partial \varphi} = -\lambda R \sin \varphi. \qquad (6.88)$$

The fundamental quantities

$$
\left.
\begin{aligned}
E &= \left(\frac{\partial X}{\partial \varphi}\right)^2 + \left(\frac{\partial Y}{\partial \varphi}\right)^2 = R^2(1 + \lambda^2 \sin^2 \varphi) \\[2mm]
F &= \frac{\partial X}{\partial \varphi}\frac{\partial X}{\partial \lambda} + \frac{\partial Y}{\partial \varphi}\frac{\partial Y}{\partial \lambda} = -\lambda R^2 \sin \varphi \cos \varphi \\[2mm]
G &= \left(\frac{\partial X}{\partial \lambda}\right)^2 + \left(\frac{\partial Y}{\partial \lambda}\right)^2 = R^2 \cos^2 \varphi.
\end{aligned}
\right\} \qquad (6.89)
$$

Since $e = R^2$ and $g = R^2 \cos^2\varphi$ it is found that

$$
\left.
\begin{aligned}
m_\varphi &= \sqrt{1 + \sin^2 \varphi} \\[2mm]
m_\lambda &= 1.
\end{aligned}
\right\} \qquad (6.90)
$$

The parallel circles are straight lines, the meridians become sinusoidal curves.
On the equator there is conformality since $m_\lambda = m_\varphi = 1$. The meridians are
orthogonal only to the equator since for $\varphi = 0$, $F = 0$; but the central meridian
is at right angles to all parallels. The general shape of the parametric network is
shown in Fig. 6.8.

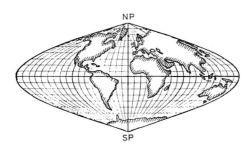

Fig. 6.8
Sanson–Flamsteed's equivalent projection

The Werner projection has the pole as a central point with $\varphi_0 = 90°$ and $\rho_0 = R \cot \varphi_0 = 0$. Then

$$\left. \begin{array}{l} \rho = \displaystyle\int_\varphi^{\varphi_{90}} R \, \mathrm{d}\varphi = R(90-\varphi) \\[2em] \vartheta = \dfrac{\lambda \cos \varphi}{90-\varphi}. \end{array} \right\} \tag{6.91}$$

and

The derivatives are

$$\left. \begin{array}{l} \dfrac{\partial \rho}{\partial \varphi} = -R; \qquad \dfrac{\partial \rho}{\partial \lambda} = 0 \\[2em] \dfrac{\partial \vartheta}{\partial \varphi} = \dfrac{-\lambda}{(90-\varphi)} \left\{ \sin \varphi - \dfrac{\cos \varphi}{(90-\varphi)} \right\} \\[2em] \dfrac{\partial \vartheta}{\partial \lambda} = \dfrac{\cos \varphi}{(90-\varphi)}. \end{array} \right\} \tag{6.92}$$

The scale distortion is

$$m_\varphi = \sqrt{1 + \lambda^2 \left\{ \sin \varphi - \dfrac{\cos \varphi}{(90-\varphi)} \right\}^2} \tag{6.93}$$

$$m_\lambda = 1.$$

Further particulars can be derived in the usual manner. The network is shown in Figure 6.9.

Fig. 6.9
Werner's equivalent projection

## Chapter 7

APPLICATION OF MAP PROJECTIONS

The various map projections discussed in previous chapters are applicable in two principal forms:

(a) As a basis for graphical representation of the terrain in maps.
(b) As a basis for digital representation of the terrain as a set of discrete points.

## 7.1  Map reference systems

In order to relate the various features depicted in maps to a common framework, such a framework has to represent a well defined coordinate system. The position of points on the terrestrial or lunar surface is uniquely defined by their curvilinear geographical coordinates, which may be either spherical or ellipsoidal latitude and longitude. In a projection system the image of such surface points is defined with the aid of plane rectangular coordinates $X$ and $Y$. The transformation from the datum surface to the projection surface takes place in accordance with the mapping equations for the chosen projection, which express the functional relationship between the two surfaces.

In practice, the transformation computation is usually limited to the intersections of a suitable densified geographical grid, i.e. to the intersections of meridians and parallels. By connecting the transformed geographical grid-points images a cartographic network is obtained, specific to the map projection utilized. This network is the basis for map construction by manual, optical-mechanical or mechanical compilation. The conventional cartographic process utilizes data obtained by measurements related to the general framework, the character and form of presentation depending on the purpose of the map being produced.

The terrain may sometimes be represented by a digital rather than a graphical model, especially in cases where the user is interested in certain selected surface features.

The electronic computer makes the mathematical transformation of points from the datum surface to the projection surface, a relatively easy task, since once the mapping equations are programmed for processing, large volumes of data may be speedily handled on a routine basis.


## 7.2   Digital terrain model and data banks

The direct and indirect measurements performed for the purpose of recording the physical and artificial (man-made) features are processed in the form of a digital model. Each point measured is defined by its three dimensional coordinates within a specific system, such as longitude, latitude and height above mean sea level or space rectangular $X$, $Y$ and $Z$ coordinates. The discrete points determined by measurements and expressed in the form of coordinates constitute a digital terrain model, which may be stored on magnetic tape, disc or a similar medium and plotted at will when a graphical presentation is required. Digital terrain models are usually created for specific purposes, such as highway design, irrigation projects etc. They carry the distinct advantage of compact storage combined with the ability to process in the computer large volume of data involved in engineering design, often without any need to resort to the slow and costly procedure of preparing a map or extracting pertinent data from it.

One can imagine a data bank as a storage divided into surface compartments, the compartment boundaries corresponding to a system of parametric lines created by the geographical or plane rectangular coordinate grid. As the various surveys progress, their digital or digitized results are entered into appropriate compartments, building up a digital terrain model of the country or the region, which maintains the data bank.

Ideally, if the area in question were covered by a very dense network of points of known elevation (say 1 m × 1 m) and a complete survey of man-made features, a map at almost any scale could be prepared from such a data bank by a suitable generalization, performed by a computer according to density of data required for the scale desired. In practice, the density of data available in any country varies fairly sharply in different regions; the survey data is seldom uniformly revised for the purpose of updating; and the digital terrain model for the whole country which the data bank is supposed to create and update, is far from ideal. It is however almost imperative to create a data bank, since this is the best way to arrive at optimum efficiency in multiscale mapping, avoid costly duplications and at the same time facilitate data processing with the aid of the electronic computer.

Map projections play an important part in such a system by providing the constantly required transformation of data from geographical to plane rectangular coordinates and vice versa, evaluating the distortions involved and providing pertinent data for the desired graphical displays.

## 7.3  Evaluation of map projections

A map projection's suitability for cartographic or geodetic purposes is established on the basis of the evaluation and analysis of the distortions caused by the projection.

The various distortions to be analysed have to be computed at a large number of points within the area under consideration and this is a task where the high-speed computer is eminently applicable. The computer output can be converted from digital to graphical form by an automatically plotted display of distortions in the form of the Tissot indicatrix.

Considering that the computation of distortions involves the use of mapping equations, the problem of automatic plotting or scribing of grids and graticules can be simultaneously solved. A generalized system, utilizing the electronic computer and an automatic plotter, was developed by Adler et al. [3] for the purpose of providing a basis for the rational choice of the map projection best suited for the purpose in hand.

The existing map projections can be compiled in a directory offering an imposing array of possibilities without seeking new projections which may be a fascinating task if an impractical one. From such a directory a rational choice can be made on the basis of scientific evaluation, as postulated strongly by Robinson [18].

Theoretically, the directory may include a comprehensive listing of existing projections such as that of Maling [14]. In practice however a selection can be made from a limited listing of, say, approximately 15 projections belonging to those most utilized for cartographic or geodetic purposes.

The directory would list the projection's name, its mapping equations

$$X = X(\varphi, \lambda)$$
$$Y = Y(\varphi, \lambda) \tag{7.1}$$

and their partial derivatives

$$\frac{\partial X}{\partial \varphi} ; \quad \frac{\partial X}{\partial \lambda} ; \quad \frac{\partial Y}{\partial \varphi} ; \quad \frac{\partial Y}{\partial \lambda}$$

stored on punched cards, magnetic tape or disc.

Referred to the $X$ and $Y$ axes of the projection's rectangular coordinate system, a vector in the positive ($\varphi$ increasing) direction of the projection of the meridian onto the map grid is:

$$e_m = \frac{1}{M} \begin{bmatrix} \dfrac{\partial Y}{\partial \varphi} \\[2ex] \dfrac{\partial X}{\partial \varphi} \end{bmatrix} \qquad (7.2)$$

where $M$ is the radius of curvature of the ellipsoid in the meridian. Considering that the element of length along the meridian on the ellipsoid is $ds_\lambda = M\,d\varphi$, and the element of length on the map grid is $ds = \sqrt{dY^2 + dX^2}$, the length ratio (scale) along the represented meridian is the length of this vector, i.e.

$$h = \frac{ds}{ds_\lambda} = |e_m| = \frac{d\varphi}{dS_\lambda}\sqrt{\left(\frac{\partial Y}{\partial\varphi}\right)^2 + \left(\frac{\partial X}{\partial\varphi}\right)^2} = \frac{1}{M}\sqrt{\left(\frac{\partial Y}{\partial\varphi}\right)^2 + \left(\frac{\partial X}{\partial\varphi}\right)^2}. \qquad (7.3)$$

Similarly, a vector along the projection of the positive direction of the parallel is

$$e_p = \frac{1}{N\cos\varphi} \begin{bmatrix} \dfrac{\partial Y}{\partial \lambda} \\[2ex] \dfrac{\partial X}{\partial \lambda} \end{bmatrix} \qquad (7.4)$$

where $N$ is the radius of curvature in the prime vertical. The length of this vector is the scale along the represented parallel, i.e.

$$k = \frac{ds}{ds_\varphi} = |e_p| = \frac{1}{N\cos\varphi}\sqrt{\left(\frac{\partial Y}{\partial\lambda}\right)^2 + \left(\frac{\partial X}{\partial\lambda}\right)^2}. \qquad (7.5)$$

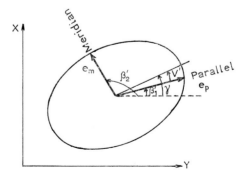

Fig. 7.1

The orientation of $e_m$ and $e_p$ vectors with respect to the $Y$-axis of the plane rectangular coordinate system of the projection

The angles made by the positive directions of the projections of the meridian and the parallel with the $Y$-axis are denoted $\beta'_2$ and $\beta'_1$ respectively (Fig. 7.1) and are given by the directions of $e_m$ and $e_p$, i.e.

$$\left. \begin{array}{l} \beta'_1 = \text{arc tan} \left( \dfrac{\dfrac{\partial X}{\partial \lambda}}{\dfrac{\partial Y}{\partial \lambda}} \right) \\[4em] \beta'_2 = \text{arc tan} \left( \dfrac{\dfrac{\partial X}{\partial \varphi}}{\dfrac{\partial Y}{\partial \varphi}} \right) \end{array} \right\} \tag{7.6}$$

with quadrant test. The angle between the projections of the meridian and the parallel is

$$\omega' = \beta'_2 - \beta'_1. \tag{7.7}$$

Now the lengths of the indicatrix semi axes $m_0$ and $m_{90}$ are computed by the usual procedure, noting that the angle between meridians and parallels in the system of geographic coordinates $\omega = 90°$.

The orientation of the indicatrix

$$\begin{array}{l} m_0 \cos V = k \cos V' \\ m_{90} \sin V = k \sin V' \end{array} \tag{7.8}$$

which eventually yields

$$\tan V' = \pm \sqrt{\frac{1 - k^2/m_0^2}{k^2/m_{90}^2 - 1}}. \tag{7.9}$$

Then $\gamma$, denoting the angle between the major axis of the indicatrix and the $Y$ axis, is given by

$$\gamma = \beta' + V'.$$

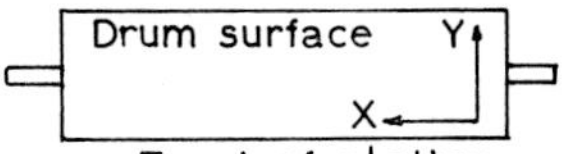

Fig. 7.2
The relationship of the projection's $X, Y$ rectangular coordinate system to the drum plotter

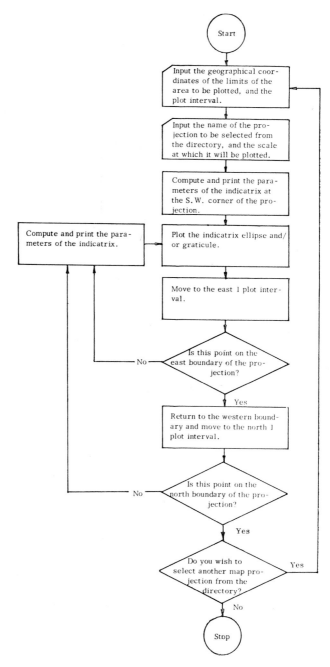

Fig. 7.3
A simplified flow chart of the computer program

Thus the parameters of the indicatrix $m_0$, $m_{90}$ and the orientation angle are completely determined for the system from the two vectors $e_m$ and $e_p$.*

All that remains is to relate the $X$, $Y$ system of projection's coordinates to the plotter.

Using, say, a drum plotter, the relationship may be illustrated schematically as in Figure 7.2.

A generalized flow chart for the computer program involved in the approach outlined is given in Figure 7.3.

Sample plots of three different conic projections generated on the IBM 1627 drum plotter are illustrated in Figures 7.4, 7.5, and 7.6. The original plotting scale was 1: 30,000,000 and the illustrations are unretouched photographic reductions of the actual plots.

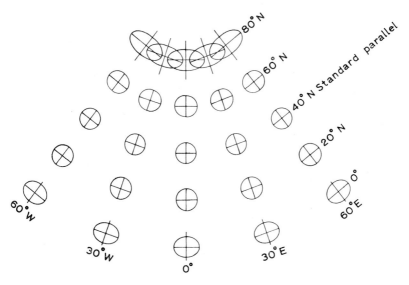

Fig. 7.4
True conic projection based on 40° N standard parallel

The evaluation of map projections with a view to making the rational choice of the most suitable projection for the purpose in hand concerns the geodesist, the geographer and the cartographer, each from his own point of view. The evaluation system based on a computer processed computation of distortions offers great flexibility, working from a directory listing of selected projections covering a wide range of possible applications.

---

* Refer also to Section 3.5, where a slightly different notation has been used.

The geodesist is interested primarily in those projections which hold the distortions to a minimum over the area in question, and in this case the directory will probably be limited to conformal projections such as Transverse Mercator (Gauss–Kruger), Universal Transverse Mercator, Lambert and perhaps for small narrow areas Cassini–Soldner. The geographer is likely to require a much wider range of projections to obtain the most desirable effect in his many kinds of presentations.

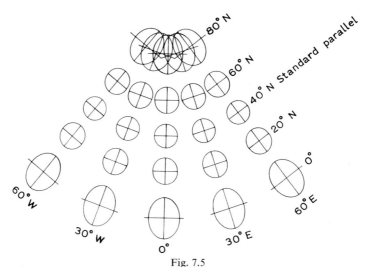

Fig. 7.5
Conic central perspective projection based on 40° N standard parallel

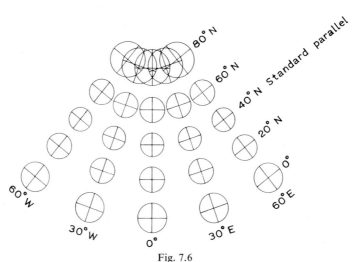

Fig. 7.6
Conic conformal projection based on 40° N standard parallel

Very often a purposely distorted map may provide a more realistic expression than a conventional one. The distortional effect can often best be examined by sample plotting of the outline of the area to be presented to obtain a better idea of the graphical aspect of the desired presentation, which is sometimes difficult to judge from the computer output.

The cartographer, besides the considerations of choice and deformational analysis is undoubtedly interested in the various aspects of automatic plotting, scribing or photo-engraving of both graticules and details. The evaluation system lays a firm basis for automatic plotting, particularly on flat-bed coordinatographs.

## 7.4  Automatic plotting

The process of depicting the spherical or spheroidal surface of the earth on a two-dimensional map involves the use of mapping equations, which are particular to the map projection chosen for the purpose.

The natural and man-made features which are expressed in the map, in most cases surveyed from aerial photographs, are referred to a framework of geodetic control. It is thus necessary to plot the graticule of the geographical system of meridians and parallels within the plane rectangular system of the projection, together with the geodetic control points.

This task is particularly suitable for automatic flat-bed coordinatographs, manufactured today by several North American and European companies.

The computation of plane rectangular coordinates from geographical coordinates is entrusted to the computer, capable of processing very large number of points quickly and efficiently. The output of this computation becomes the input for the automatic coordinatograph, which must be capable of automatically drawing, pricking, engraving or photo exposing with uniform, specified considerable precision. Three basic units comprise the automatic plotting system: (a) An input-output device, (b) an electronic control system, (c) a plotting table with accessories.

The modern systems have facilities for the following operations:

(1) The pricking (scoring) of discrete points to a preselected depth on various map materials. In most cases the scored points can be marked and identified by symbols and numbers produced with the aid of a printing or photo-exposing device.

The positioning accuracy of the operating tool on the table is of the order $\pm 0.02$ mm $- \pm 0.05$ mm which is certainly well within the graphical tolerances required in cartography.

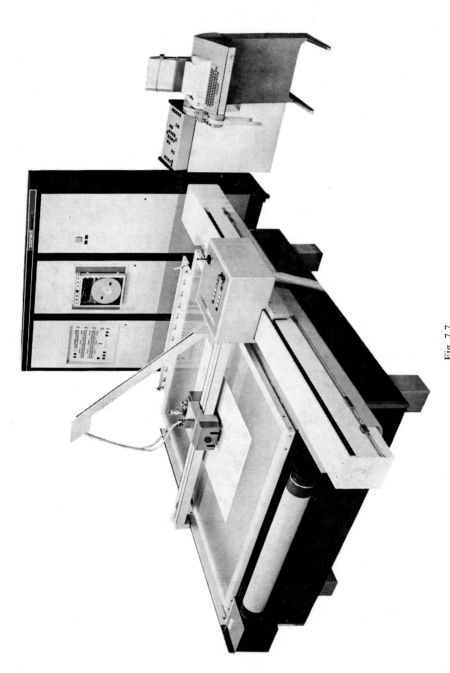

Fig. 7.7

Coradomat (courtesy of Coradi, Switzerland)

(2) The drawing of maps and plans by means of straight and curved lines of varying thicknesses. This may be accomplished by pencil, ball point pen or rapidograph.

The drawing tools are mounted in a turret in such a way that the desired tool is automatically selected according to computer prepared instructions.

The positioning and plotting are controlled electronically by a suitably coded input.

Maps may be engraved (scribed) on coated foils by using sapphire scribers instead of drawing tools.

Some of the latest automatic flat-bed plotters are equipped with lightscribing and lightprinting devices exposing at varying line thicknesses in combination with light projection of figures, letters and symbols. In this type of operation the coordinatograph must operate in photo-darkroom conditions.

(3) The digitizing of existing maps and plans, which in essence is a process reverse to plotting. The existing graphical material is converted into a digital terrain model, which may either serve the purpose of creating a data bank or be reprocessed in the computer for different forms of desired presentations either digital or graphical.

Figures 7.7 and 7.8 illustrate two of the several models of automatic coordinatographs currently available on the market for cartographic purposes.

Fig. 7.8
Gerber model 2032B Automatic Drafting System (courtesy of Gerber Scientific Instrument Cy, Hartford, Conn.)

## 7.5    Universal Transverse Mercator (UTM) projection system

It has been a generally accepted natural development that each country chose a map projection best suited for its geodetic and cartographic purposes considering the size and shape of the area to be represented and the characteristics most desirable in the representation. In many cases the basic projections were modified to serve such purposes. Obviously no single map projection could be universally accepted for world-wide application, not only because of the fact that there is no such ideal multi-purpose projection, but also because, more often than not, the national considerations override the international ones.

Thus the world geodetic and cartographic community has learned to live with the co-existence of a rather large number of map projections, some being more popular than others, and often several projections being utilized overlappingly in the same area.

During World War II the need for a world wide plane coordinate system was brought up by the military, who specified the criteria for such a reference system as follows:

(1) Conformality in order to minimize directional errors.

(2) "Continuity" over sizeable areas coupled with a minimum number of zones.

(3) Scale errors caused by the projection not to exceed a specified tolerance.

(4) Unique referencing in a plane rectangular system of coordinates for all zones.

(5) Transformation formulae from one zone to another to be uniform throughout the system (assuming one reference ellipsoid).

(6) Meridional convergence not to exceed five degrees.

On the basis of the above mentioned criteria, the Universal Transverse Mercator system, commonly known as the UTM, was developed for world-wide application. It is, in essence, a modification of the Transverse Mercator projection, sometimes referred to as Gauss-Krüger projection.

The basic features of the UTM are as follows:

(1) The world is divided into 60 zones, each extending through 6° of longitude. The zones are numbered consecutively from one to sixty beginning with the zone between 180° W and 174° W and continuing eastwards. The zone numbering system is in accordance with the agreement made in connection with the International Map of the World at the 1: 1,000,000 scale. In the geometrical sense the transverse cylinder is tangent along the central meridian of each zone (assuming the datum surface being a sphere). Thus the zone extends 3° Eastwards and 3° Westwards of the central meridian, which is of course a great circle.

The same division applies in the ordinary Transverse Mercator used by the U.S.S.R. and sometimes known as the Soviet Unified Reference System.

(2) A scale distortion or grid scale constant of 0.9996 is applied along the central meridian of each zone to improve scale retention characteristics of the projection. This is the principal feature of the UTM and should be carefully considered.

In the ordinary Transverse Mercator the central meridian is a standard great circle along which there is no scale distortion, i.e. the scale factor $m_0$ equals 1.0000. The small circles parallel to it are represented by the vertical lines of the rectangular grid and thus there is a scale increase away from the central meridian. Assuming a zone width of 6°, or approximately 350 km Eastwards and Westwards of the central meridian, the scale factors throughout the zone are illustrated in Figure 7.9.

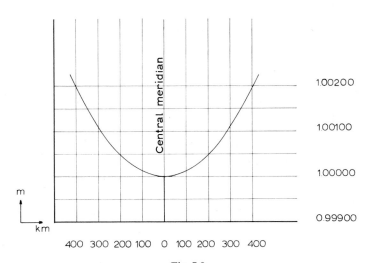

Fig. 7.9
Deterioration of scale as a function of distance from the central meridian in the Transverse Mercator projection

By assigning the scale factor $m_0$ of 0.9996 to the central meridian in the UTM system, the effect is that of the transverse cylinder becoming secant to the datum surface instead of tangent. Instead of the one standard meridian we now have two, along the lines of "secancy", and the scale distortions are more favourably spread out over the zone – see Figure 7.10 (see Section 5.1.4).

The choice of $m_0 = 0.9996$ for the central meridian was made so as to limit the scale error to $\frac{1}{2500}$ within the zone.

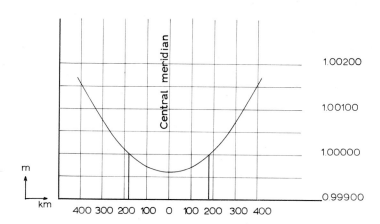

Fig. 7.10

Scale factor distribution in an UTM zone. Note that the two standard meridians are 180 km East and West of the central meridian

The approximate formula for the computation of the scale factor:

$$m_p = m_0 \left( 1 + \frac{Y'^2}{2R^2} \right) \qquad (7.10)$$

where $m_0$ = scale distortion assigned to the central meridian

$Y'$ = scaled distance on the projection from the central meridian

$R$ = $\sqrt{MN}$ = mean radius of the earth

$m_p$ = scale distortion at the point in question.

The specific range of the scale error depends also on the latitude. The simplest solution to limiting the scale error to a strict prespecified tolerance is to introduce sub-zones or to limit the zone width, but since the UTM system was designed for world-wide coverage, a very large number of zones would defeat the original purpose.

(3) A plane rectangular metric grid is superimposed on each zone, assigning a 500,000 m false Easting (Y coordinate value) to the central meridian, a 0 (zero) Northing (X coordinate) to the Equator for the Northern hemisphere and a false Northing of 10,000,000 m to the Equator for the Southern hemisphere. This simple arrangement eliminates negative coordinate values. Transformation from one zone to another is according to uniform formulae, which can be easily programmed for the computer, thus eliminating the need for auxiliary tables used in manual computation.

A grid overlap between zones is customary approximately 40 km on either side of the zone boundary.

In order to facilitate local engineering projects or for certain military applications points within the overlap appear in coordinate lists for both zones but the grid references should always be given within the actual zone.

The computations in the UTM system require definition of the following terms:

$FN = FX$ = false Northing = 0 m for the Northern hemisphere

$\qquad\qquad\qquad\qquad$ = 10,000,000 m for the Southern hemisphere

$FE = FY$ = false Easting = 500,000 m

$E' = Y'$ = grid distance in metres from the central meridian (always positive)

$E = Y$ = grid Easting ($Y$ coordinate)

$N = X$ = grid Northing ($X$ coordinate)

$\Delta\lambda$ = difference in seconds of longitude from the central meridian

$\qquad$ = $\lambda_0 - \lambda$ in the Western hemisphere

$\qquad$ = $\lambda - \lambda_0$ in the Eastern hemisphere

$p$ = 0.0001 $\Delta\lambda$

$q$ = 0.000001 $\delta E'$

$\varphi'$ = latitude of the foot of the perpendicular from the point to the central meridian

$S$ = true meridional distance on the ellipsoid from the equator

$m_0$ = central meridian projection scale factor

(I) $\quad = m_0 S$

(II) $\quad = \frac{1}{2} N \sin\varphi \cos\varphi \sin^2 1'' m_0 10^8$

(III) $\quad = \frac{1}{24} \sin^4 1'' \sin\varphi \cos^3\varphi (5 - \tan^2\varphi + 9\varepsilon'^2 \cos^2\varphi + 4\varepsilon'^4 \cos^4\varphi) m_0 10^{16}$

(IV) $\quad = N \cos\varphi \sin 1'' m_0 10^4$

(V) $\quad = \frac{1}{6} \sin^3 1'' N \cos^3\varphi (1 - \tan^2\varphi + \varepsilon'^2 \cos^2\varphi) m_0 10^{12}$

(VI) $\quad = \dfrac{\tan\varphi'}{2 N^2 \sin 1''} (1 + \varepsilon'^2 \cos^2\varphi) \dfrac{1}{m_0^2} 10^{12}$

(VII) $= \dfrac{\tan\varphi'}{24 N^4 \sin 1''} (5 + 3\tan^2\varphi + 6\varepsilon'^2 \cos^2\varphi - 6\varepsilon'^2 \sin^2\varphi - 3\varepsilon'^4 \cos^4\varphi -$

$$-9\varepsilon'^4 \cos^2\varphi \sin^2\varphi) \dfrac{1}{m_0^4} 10^{24}$$

$$\text{(VIII)} = \frac{\sec \varphi'}{N \sin 1''} \frac{1}{m_0} 10^6$$

$$\text{(IX)} = \frac{\sec \varphi'}{6N \sin 1''} (1 + 2 \tan^2 \varphi + \varepsilon'^2 \cos^2 \varphi) \frac{1}{m_0^3} 10^{18}$$

$$\text{(X)} = \sin \varphi \; 10^4$$

$$\text{(XI)} = \tfrac{1}{3} \sin^2 1'' \sin \varphi \cos^2 \varphi (1 + 3\varepsilon'^2 \cos^2 \varphi + 2\varepsilon'^4 \cos^4 \varphi) 10^{12}$$

$$\text{(XII)} = \frac{\tan \varphi'}{N \sin 1''} \frac{1}{m_0} 10^6$$

$$\text{(XIII)} = \frac{\tan \varphi'}{3N^3 \sin 1''} (1 + \tan^2 \varphi - \varepsilon'^2 \cos^2 \varphi - 2\varepsilon'^4 \cos^4 \varphi) \frac{1}{m_0^3} 10^{18}$$

$$\bar{A} = \tfrac{1}{720} p^6 \sin^6 1'' N \sin \varphi \cos^5 \varphi (61 - 58 \tan^2 \varphi + \tan^4 \varphi + 270 \varepsilon'^2 \cos^2 \varphi -$$
$$- 330 \varepsilon'^2 \sin^2 \varphi) m_0 10^{24}$$

$$\bar{B} = \tfrac{1}{120} p^5 \sin^5 1'' N \cos^5 \varphi (5 - 18 \tan^2 \varphi + \tan^4 \varphi + 14 \varepsilon'^2 \cos^2 \varphi -$$
$$- 58 \varepsilon'^2 \sin^2 \varphi) m_0 10^{20}$$

$$\bar{C} = \tfrac{1}{15} p^5 \sin^4 1'' \sin \varphi \cos^4 \varphi (2 - \tan^2 \varphi) 10^{20}$$

$$\bar{D} = q^6 \frac{\tan \varphi'}{720 N^6 \sin 1''} (61 + 90 \tan^2 \varphi + 45 \tan^4 \varphi + 107 \varepsilon'^2 \cos^2 \varphi$$
$$- 162 \varepsilon'^2 \sin^2 \varphi - 45 \varepsilon'^2 \tan^2 \varphi \sin^2 \varphi) \frac{1}{m_0^6} 10^{36}$$

$$\bar{E} = q^5 \frac{\sec \varphi'}{120 N^5 \sin 1''} (5 + 28 \tan^2 \varphi + 24 \tan^4 \varphi + 6 \varepsilon'^2 \cos^2 \varphi +$$
$$+ 8 \varepsilon'^2 \sin^2 \varphi) \frac{1}{m_0^5} 10^{30}$$

$$\bar{F} = q^5 \frac{\tan \varphi'}{15 N^5 \sin 1''} (2 + 5 \tan^2 \varphi + 3 \tan^4 \varphi) \frac{1}{m_0^5} 10^{30}.$$

*Computation of UTM grid coordinates from geographic position*

$$N = X = \text{(I)} + \text{(II)} p^2 + \text{(III)} p^4 + \bar{A} \tag{7.11}$$

$$E = Y = FE \pm E' \tag{7.12}$$

$$E' = \text{(IV)} p + \text{(V)} p^3 + \bar{B} \tag{7.13}$$

(plus for points East and minus for points West of the central meridian)

*Geographic position from UTM grid*

$$\varphi = \varphi' - (VI)q^2 + (VII)q^4 - \bar{D} \qquad (7.14)$$

$$\varDelta\lambda = (VIII)q - (IX)q^3 + \bar{E} \qquad (7.15)$$

$$\lambda = \lambda_0 \pm \varDelta\lambda \qquad (7.16)$$

$C$ = grid convergence = $(X)p + (XI)p^3 + \bar{C}$

Grid azimuth = geodetic azimuth $\pm C$ (plus $C$ for points East and minus $C$ for points West of the central meridian).

In the past tables have been compiled for the expressions in Roman numerals and graphs for the letter terms to facilitate computation.

Using a high speed computer there is no need for tables and graphs and the individual expressions given above may be useful in programming.

Attention must be paid to the ellipsoidal parameters in the above expressions, since different ellipsoids were used in various parts of the world.

(4) The UTM system is used between latitudes of 84° N and 80° S. Both polar regions are covered by the UPS (Universal Polar Stereographic) System which complements the UTM but is quite independent of it. There is an overlap area along the boundary of the two systems.

Approximately 60 countries use the UTM as the most authoritative and general use projection within the country, even though many of the 60 use also secondary projections and grid references.

The U.S.S.R., China and other countries of the Eastern Bloc use the Transverse Mercator (Gauss–Krüger) with 6° zones.

Approximately 50 countries use other projections than UTM or Transverse Mercator with standard 6° zones. Whilst it is difficult to advocate the UTM system as the universal one which all countries should be pressed to adopt, there is much to be said for a uniform system's world wide acceptance.

## 7.6   Map projections for lunar applications

### 7.6.1   *Introduction*

Man's physical presence on the moon puts the problem of lunar surveying and mapping in a new context. It is not unreasonable to assume that more landings will follow and that eventually permanent outposts will be established on the surface of the moon. These lunar stations will probably be of limited size, their sole function being scientific study with perhaps exploitation of natural resources following at a later stage.

Certainly the establishment of astronomic observatories where the heavens can be observed free of the limitations of an atmosphere is feasible. Likewise, the collection of data on the hazards of space cosmic rays, dust, meteorities etc., preparatory to missions to other parts of the solar system, is within the anticipated scope of lunar activities.

In the period preceding the American landings, maps were prepared, based on the photographic information supplied by lunar probes, ranging in area coverage from the entire visible lunar surface to specific areas of a few square kilometres. These maps range in scale from small scale hemispheric maps to 1: 500 maps of landing areas. As more precise and more extensive data becomes available, accurate maps for the entire lunar surface must be compiled. The question of considering the most suitable map projection for lunar mapping is no longer an academic exercise, it is a professional reality. Alongside with mapping problems selenodetic control for lunar surveying tasks should be considered. A permanent lunar station with a range of operational activities of say one or two degrees of latitude and longitude requires a system of plane rectangular coordinates, much in the same way as the various states and countries on the earth. The selection of the appropriate lunar map projections must take into account the size and shape of the moon on the one hand and the prospective applications from the point of the various kinds of distortion tolerances on the other.

### 7.6.2   *Parameters of the moon*

The figure of the moon is known to be best represented by a triaxial ellipsoid. Defining:

$a$ = length of the semi-axis directed through the intersection of the lunar equator and the lunar zero meridian (close to the crater known as Mösting A).

$b$ = length of the semi-axis coincidental with the moon's mean librational axis.

$c$ = length of the semi-axis mutually perpendicular to axes $a$ and $b$,

then:

$a = 1738.57$ km
$b = 1737.49$ km
$c = 1738.21$ km.

Thus the maximum flattening of the moon:

$$f_{max} = \frac{a-b}{a} = 0.0005177204$$

and the minimum flattening:

$$f_{min} = \frac{c-b}{c} = 0.000408466.$$

Using the value for the maximum flattening:

first eccentricity $= \varepsilon^2 = 2f - f^2 = 0.0010351728$    see form. (2.3)

second eccentricity $= \varepsilon'^2 = \frac{\varepsilon^2}{1-\varepsilon^2} = 0.0013634552$    see form. (2.5).

For the purpose of investigating lunar map projections – if the moon is considered a rotational ellipsoid – the maximum flattening should be used to ensure that any errors introduced into the projection as a result of flattening will be the maximum errors for any point on the lunar surface.

When the moon is represented by a sphere, the radius of the sphere is the mean lunar radius:

$R = 1738.09$ km.

Using this radius, one degree of arc on the spherical reference surface becomes:

$$1° \text{ of lunar arc} = \frac{2\pi}{360} \times 1738090 \text{ m} = 30335 \text{ m.}$$

### 7.6.3   Azimuthal lunar projections

The azimuthal projections should be considered for applications over limited areas for large scale lunar cartography or surveying operations close to a permanent station. As pointed out earlier (Ch. 4) these projections in the geometrical sense assume the projection plane surface being tangent to the datum surface at a single point. It should be remembered that the true scale (scale factor $m = 1.0$) is preserved only at this point. Furthermore angles measured on the projection surface at the point of tangency are undistorted, i.e. corresponding to angles measured on the datum surface. If this point of tangency were made to coincide with say a lunar base or a landing site, the azimuth from such a station would be accurately represented on the projection. Using the base as a center of operations for local investigations this property would be valuable to the astronauts manning the station.

On the earth, one of the azimuthal projections, the stereographic, is commonly used in conjunction with the UTM System to represent polar areas.

Should the Transverse Mercator or the UTM projections be used for representation of the entire lunar surface, the stereographic projection would again be the obvious choice for the polar areas of the moon.

The U.S. Army Map Service (now TOPOCOM) has used a perspective azimuthal projection, referred to as the AMS Lunar Projection. In this projection the perspective center lies on the extension of the radius vector defined by the center of the datum surface and the point of tangency of the projection plane. The distance of the perspective center from the center of the moon in the AMS lunar projection is $D = 1.53748\,R$, where $R$ is the mean radius of the moon (see Fig. 7.11).

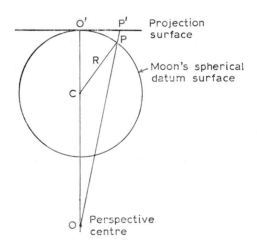

Fig. 7.11
AMS lunar projection ($OC = D = 1.53748\,R$)

The mapping equations for the AMS Lunar projection are as follows:

$$X = \frac{R(D+R)(\cos \varphi_0 \sin \varphi - \sin \varphi_0 \cos \varphi \cos \lambda)}{D+R(\sin \varphi_0 \sin \varphi + \cos \varphi_0 \cos \varphi \cos \lambda)} \tag{7.17}$$

$$Y = \frac{R(D+R)\cos \varphi \sin \lambda}{D+R(\sin \varphi_0 \sin \varphi + \cos \varphi_0 \cos \varphi \cos \lambda)}, \tag{7.18}$$

where $\lambda_0$, $\varphi_0$ are the lunar longitude and latitude of the projection center (point
of tangency) respectively, $\lambda$, $\varphi$ are the lunar longitude and latitude of
the point being mapped.

$D$ is the distance of the perspective point from the center of the moon.

$R$ is the mean lunar radius.

The only clearly stated requirement taken into consideration was that the
graticule should extend approximately $100°$ from the origin near the center of
the moon's visible disc to accomodate those marginal parts, which are perio-
dically visible from the Earth as the result of the physical and optical librations
of the moon. Maling has discussed the suitability of lunar projections in [13]
already some years ago when the lunar exploration was at a relatively unad-
vanced stage.

Whilst the merits of many varieties of azimuthal projections should not be
underestimated for local purposes of polar areas, the choice of an azimuthal
projection for the entire lunar surface would not serve today's objectives in
lunar cartography and selenodetic control.

### 7.6.4  Cylindrical conformal lunar projections

The conditions of conformality and equidistancy are the most desirable in
analysing the various available projections with a view to their lunar applications. It is obviously an impossibility to satisfy both of those conditions at all
points on the ellipsoidal or spherical datum surface.

It is therefore logical to seek conformality without serious length distortions.
As stated earlier the directional or angular errors in lunar projection systems
are less desirable than errors in length representation.

The Transverse Mercator and the UTM projection therefore seem to be the
obvious choices for a system to represent the entire lunar surface, as they do on
Earth, supported by the Stereographic projection for the polar areas. The
continuity of the system is preserved, which is a most important factor for a
"universal" representation.

The width of the Transverse Mercator lunar zones and the central meridian
scale factor in the lunar UTM have to be carefully considered.

Myers [17] carried out recently an evaluation, under the guidance of the
authors, of the feasible lunar projections, utilizing the computer system established by Adler et al. [3], in order to obtain a comparison of the distortional
characteristics.

In its preliminary stage the evaluation considered the sensitivity of the scale
to the flattening of the reference surface. Assuming six degree zones in the

Transverse Mercator and the UTM, the maximum effect of the flattening

$$dm = m_0 (0.0027 \, df) \tag{7.19}$$

which meant that the scale changes only very slightly for large variations in flattening. Changing the previously mentioned flattening of the moon $f = 0.0005177204$ to a flattening of zero, i.e. a sphere, yields a change in scale ($dm$) of approximately one part in a million, which means that the lunar datum surface can be considered to be a sphere for the Transverse Mercator as far as the scale errors are concerned.

### 7.6.5 *Comparison of distortional characteristics of lunar projections*

Let us first consider three azimuthal projections: the AMS Lunar, the stereographic and the orthographic. These projections were evaluated for directional and linear distortions within a 60 km ($2°$ of arc on lunar surface) radius and within a 300 km ($10°$ of arc) radius from the point of origin. The first radius represents the range of operations from a lunar station, the second one is taken with view to representing polar areas of the moon in conjunction with a Transverse Mercator projection.

The results are presented in Table 7.1.

TABLE 7.1

*Angular and linear distortions of lunar azimuthal projections as a function of distance from the origin*

| Azimuthal Projection | Angular distortion in minutes of arc | | Linear distortion of a 30 km line in parts in 10,000 | | | | |
|---|---|---|---|---|---|---|---|
| | | | With no scale factor* | | 0.994 scale factor* applied | | |
| | At 60 km from origin | At 300 km from origin | At 60 km from origin | At 300 km from origin | At origin | At 60 km from origin | At 300 km from origin |
| Stereo-graphic | 0 | 0 | 3 | 77 | 30 | 27 | 46 |
| AMS Lunar | 0.11 | 5.57 | 3 | 60 | 24 | 23 | 36 |
| Ortho-graphic | 2.40 | 26.3 | 14 | 152 | 0 | 14 | 152 |

\* Projection scale factor $m_0$.

The orthographic projection is obviously the least desirable of the three and must be rejected. There is little to choose between the other two as far as the linear distortions are considered. The application of a 0.994 scale factor at the point of tangency (origin) produces the secant effect and would be useful. Favour may be given in the final choice to the Stereographic because of its conformality.

Next, the Transverse Mercator and the UTM projections were compared for linear distortions, at various lunar latitudes and as a function of distance from the central meridian within a 6° zone. There is of course no directional distortion at a point since the projections are conformal.

The results are given in Table 7.2.

TABLE 7.2

*Scale distortion on the Transverse Mercator and UTM projections as function of lunar latitude and distance from central meridian*

| $\varphi$ | $d\lambda$ | Transverse Mercator scale factor | UTM scale factor |
|------|------|------|------|
| 0°  | 0°   | 1.000,000 | 0.999,600 |
|     | 1.5° | 1.000,410 | 1.000,010 |
|     | 3°   | 1.000,770 | 1.000,369 |
| 40° | 0°   | 1.000,000 | 0.999,600 |
|     | 1.5° | 1.000,601 | 1.000,201 |
|     | 3°   | 1.001,206 | 1.000,805 |
| 80° | 0°   | 1 000,000 | 0.999,600 |
|     | 1.5° | 1.000,744 | 1.000,343 |
|     | 3°   | 1.001,776 | 1.001,373 |

With six degree zones the UTM is preferable slightly to the Transverse Mercator as it is on the Earth. Either projection will adequately fullfil the general moon mapping requirements. Consideration should be given to the use of three degree zones, the implication of increasing the number of zones being much less serious than on the Earth in view of limited character of lunar surface activity.

### 7.6.6  *Lunar map projections – summary*

1. A spherical datum surface may be assumed for the moon, the flattening being so slight that the approximation of the figure of the moon by a triaxial ellipsoid would produce no significant improvement.

2. In view of the global character of moon mapping, the Universal Transverse Mercator or the Transverse Mercator system with six or three degree zones appears to be the best projection for this purpose.

3. The Stereographic projection may be considered as the most suitable for mapping and surveying of small areas of lunar surface, considered as permanent moon stations. The same projection could be used for the lunar polar regions in conjunction with the lunar UTM or Transverse Mercator system. A projection scale factor at the point of tangency could be introduced to improve linear distortion characteristics.

4. Other lunar projections than those mentioned above may be useful for special presentations, where certain distortional effects may even be desirable. It is however highly improbable that they would be found suitable for the purpose of lunar cartography and selenodesy.

APPENDIX A

CONFORMAL PROJECTIONS

DIFFERENCES BETWEEN ELLIPSOIDAL AND PLANE DISTANCES;
ARC TO CHORD CORRECTION; MERIDIAN CONVERGENCE

## A.1 Introduction

Although angles are transformed undistorted it remains in many instances necessary to calculate corrections to ellipsoidal elements in order to obtain corresponding elements in the map, and vice versa. The plane situation is shown in Figure A.1.

Suppose two points $p$ and $q$ on the ellipsoid are projected in the plane as $P$ and $Q$. The shortest connection on the ellipsoid is the geodesic $p\,q$ of a length $s$. This curve however, does not as a rule project as a straight line $PQ$, but transforms into the curve $PQ$ of a length $S$. This length is first to be expressed in

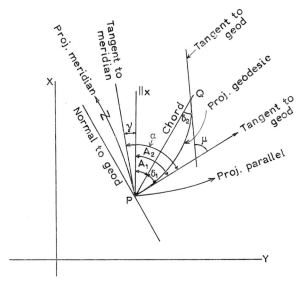

Fig. A.1
Plane projection of the geodesic $PQ$

151

terms of $s$ and the scale distortion $m$. Furthermore the curvature of $PQ$ is needed in terms of $S$ and $m$.

This relationship has first been derived by Schols [21]. Then the length correction, giving the difference between the chord length $D$, and $S$ of the projected geodesic can be calculated, as well as the angle between the tangent at $P$ to $PQ$ and the chord $PQ$.

This angle is denoted by $\delta_1$. The corresponding angle at $Q$ is $\delta_2$. If the bearing of $PQ$ and the tangent with respect to the $X$-axis are indicated by $A_1$ and $A_2$ respectively, then

$$\delta_1 = A_2 - A_1 \qquad (A.1)$$

This angle is usually called the "arc to chord" correction. Sometimes in the literature the bearings $A_1$ and $A_2$ are denoted by "$t$" and "$T$" respectively the correction being named "$T-t$" correction. In this book the former nomenclature is adhered to.

The inverse formulae are also given. In addition the deviation $\Delta$ of the curve $PQ$ from the chord is derived. Finally the convergence $\gamma$ in the projection is determined i.e. the angle at $P$ between the $X$ abscissa direction and the projected meridian. It will be shown that at first approximation this convergence is independent of the projection viz.

$$\gamma = \lambda \sin \varphi \ .$$

## A.2    The length of the projected geodesic

The length of the projected geodesic $PQ$ is determined in terms of the ellipsoidal length $S$ and scale distortion $m$.

Considering that

$$m = \frac{\mathrm{d}S}{\mathrm{d}s}$$

then

$$S = \int_0^s m \, \mathrm{d}s \ . \qquad (A.2)$$

Expanding $m$ in terms of $s$ leads to the general formula

$$m = m_P + \left(\frac{\partial m}{\partial s}\right)_P s + \frac{1}{2}\left(\frac{\partial^2 m}{\partial s^2}\right)_P s^2 + \frac{1}{6}\left(\frac{\partial^3 m}{\partial s^3}\right)_P s^3 + \dots \qquad (A.3)$$

($m_p$ denotes the scale distortion at $P$, the zero point of the arc).

It is preferable, however, to express $m$ as a function of $S$, that is in fact in terms of the plane coordinates $X$ and $Y$:

$$\frac{\partial m}{\partial s} = \frac{dm}{dS} \cdot \frac{dS}{ds} = m \frac{dm}{dS} = mm' \tag{A.4}$$

$$\frac{\partial^2 m}{\partial s^2} = m^2 \frac{d^2 m}{dS^2} + m \left(\frac{dm}{dS}\right)^2 = m^2 m'' + mm'^2. \tag{A.5}$$

Hence (A.3) becomes

$$m = m_P + m_P m'_P s + \tfrac{1}{2}(m^2 m''_P + m_P m'^2_P)s^2 + \dots$$

and

$$S = \int_0^s m \, ds = m_P s + \tfrac{1}{2} m_P m'_P s^2 + \tfrac{1}{6}(m_P^2 m''_P + m_P m'^2) s^3 + \dots \tag{A.6}$$

This formula may be simplified if the third order terms are eliminated by the introduction of the scale distortion $m_{\frac{1}{3}}$ with reference to a point at $\tfrac{1}{3} PQ$ viz. $m_{\frac{1}{3}}$ at $\tfrac{1}{3}$ arc $S$.

By differentiation of (A.3) and using (A.4) it is seen that

$$m_{\frac{1}{3}} m'_{\frac{1}{3}} = m_P m'_P + \tfrac{1}{3}(m_P^2 m''_P + m_P m'^2_P)s + \dots$$

and

$$m_P m'_P = m_{\frac{1}{3}} m'_{\frac{1}{3}} - \tfrac{1}{3}(m_P^2 m''_P + m_P m'^2_P)s + \dots \tag{A.7}$$

Hence (A.6) becomes

$$S = m_P s + \tfrac{1}{2} m_{\frac{1}{3}} m'_{\frac{1}{3}} s^2 + \text{terms of 4}^{\text{th}} \text{ order etc.} \tag{A.8}$$

The occurence of a different scale distortion and its derivatives (at $P$ and at $\tfrac{1}{3} S$) may be considered a slight disadvantage.

It may be overcome by using the scale distortion $m_{\frac{1}{2}}$ at $\tfrac{1}{2} S$.

Then

$$S = \int_{-\frac{1}{2}s}^{+\frac{1}{2}s} \{m_{\frac{1}{2}} + m_{\frac{1}{2}} m'_{\frac{1}{2}} s + \tfrac{1}{2}(m_{\frac{1}{2}}^2 m''_{\frac{1}{2}} + m_{\frac{1}{2}} m'^2_{\frac{1}{2}})s^2 + \dots\} \, ds$$

so that

$$S = m_{\frac{1}{2}} s + \frac{1}{24}(m_{\frac{1}{2}}^2 m''_{\frac{1}{2}} + m_{\frac{1}{2}} m'^2_{\frac{1}{2}})s^3 + \dots \tag{A.9}$$

In this case the second order terms have been eliminated. The length $s$ of the geodesic on the ellipsoid in terms of $S$, its plane projection, is calculated by a direct inversion of the series (A.6), (A.8) or (A.9). It may also be done by expanding

$$s = \int_0^s \frac{1}{m} \, dS = \left(\frac{1}{m}\right)_P s + \frac{1}{2}\left(\frac{1}{m}\right)_P' s^2 + \frac{1}{6}\left(\frac{1}{m}\right)_P'' s^3 + \ldots \tag{A.10}$$

Without further derivations, the following formula is given

$$s = \left(\frac{1}{m}\right)_{\frac{1}{2}} S + \frac{1}{24}\left(\frac{1}{m}\right)_{\frac{1}{2}}'' S^3 + \frac{1}{1920}\left(\frac{1}{m}\right)_{\frac{1}{2}}'''' S^5 + \ldots \tag{A.11}$$

## A.3   The curvature of the projected geodesic. Schols' formula

The curvature of the conformally projected geodesic $PQ$ in Figure A.1 is given by Schols' formula [21]:

$$\frac{1}{\rho_{PQ}} = \frac{1}{m} \frac{dm}{dS_{(A_2-90°)}} = \frac{d \ln m}{dS_{(A_2-90°)}} \tag{A.12}$$

where

$\rho_{PQ}$ is the radius of curvature of the projected geodesic;

$S$ is the length of the curve $PQ$ in the projection plane;

d ln $m$ is the rate of change of ln $m$ per unit length d$S$ in the direction $(A_2-90°)$ of the normal to the curve at $P$.

There are several proofs of this formula. A very direct one is given by Hotine [11] and related here:

According to elementary differential geometry the curvature is equal to

$$\frac{1}{\rho} = \frac{d\mu}{dS} \tag{A.13}$$

where the tangent moves through an angle d$\mu$ following the curve over a distance d$S$.

The bearing $A_2$ of the geodesic about the $X$-axis is (see Fig. A.1)

$$A_2 = \alpha - \gamma \tag{A.14}$$

where $\alpha$ is the azimuth of the geodesic $PQ$ about the projected meridian and $\gamma$ the convergence.

Hence

$$dA_2 = d\alpha - d\gamma = d\mu . \tag{A.15}$$

It is known by Clairaut's theorem that at a point $P$ of a geodesic on a surface of revolution, the product of the radius of the parallel circle of $P$ and the sine of the azimuth of the geodesic at that point is constant:

$$N \cos \varphi \sin \alpha = c. \tag{A.16}$$

By differentiation of this equation with respect to $\varphi$

$$\frac{d(N \cos \varphi)}{d\varphi} \sin \alpha + N \cos \varphi \cos \alpha \frac{d\alpha}{d\varphi} = 0$$

or since

$$\frac{\partial N \cos \varphi}{\partial \varphi} = -M \sin \varphi,$$

$$-M \sin \varphi \sin \alpha + N \cos \varphi \cos \alpha \frac{d\alpha}{d\varphi} = 0$$

or

$$N \cos \varphi \cot g \, \alpha \frac{d\alpha}{d\varphi} = M \sin \varphi. \tag{A.17}$$

According to (3.13) and (3.16)

$$\cot \alpha = \sqrt{\frac{e}{g}} \frac{d\varphi}{d\lambda} = \frac{M}{N \cos \varphi} \frac{d\varphi}{d\lambda}$$

whence (A.17) becomes

$$\frac{d\alpha}{d\lambda} = \sin \varphi$$

and

$$d\alpha = \sin \varphi \, d\lambda \,. \tag{A.18}$$

In calculating $d\gamma$ isometric coordinates are used for convenience

$$d\gamma = \frac{\partial \gamma}{\partial \psi} d\beta + \frac{\partial \gamma}{\partial \lambda} d\lambda. \tag{A.19}$$

Within the infinitely small area about $P$ it is seen that (see Fig. A.2)

$$\left. \begin{aligned} \partial X &= mN \cos \varphi \cos (360° - \gamma) \partial \psi = mN \cos \varphi \cos \gamma \, \partial \psi \\ \partial Y &= mN \cos \varphi \sin (360° - \gamma) \partial \psi = -mN \cos \varphi \cos \gamma \, \partial \psi. \end{aligned} \right\} \tag{A.20}$$

(An element $N \cos \varphi \, d\psi$ on the meridian is projected as an element $mN \cos \varphi \, d\psi$ in the plane.)

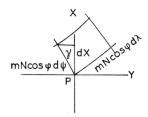

Fig. A.2
Differential projected elements

Hence in combination with (5.13)

$$\left. \begin{array}{l} \dfrac{\partial X}{\partial \psi} = \dfrac{\partial Y}{\partial \lambda} = mN \cos \varphi \cos \gamma \\[4mm] \dfrac{\partial X}{\partial \lambda} = -\dfrac{\partial Y}{\partial \psi} = mN \cos \varphi \sin \gamma. \end{array} \right\} \qquad (A.21)$$

By differentiation of these equations we get

$$\left. \begin{array}{l} \dfrac{\partial^2 X}{\partial \psi^2} = \dfrac{\partial^2 Y}{\partial \lambda\, \partial \psi} = \cos \gamma \dfrac{\partial m\, N \cos \varphi}{\partial \psi} - mN \cos \varphi \sin \gamma \dfrac{\partial \gamma}{\partial \psi} \\[4mm] \dfrac{\partial^2 X}{\partial \psi\, \partial \lambda} = \dfrac{\partial^2 Y}{\partial \lambda^2} = \cos \gamma \dfrac{\partial m\, N \cos \varphi}{\partial \lambda} - mN \cos \varphi \sin \gamma \dfrac{\partial \gamma}{\partial \lambda} \\[4mm] \dfrac{\partial^2 X}{\partial \lambda^2} = -\dfrac{\partial^2 Y}{\partial \lambda\, \partial \psi} = \sin \gamma \dfrac{\partial m\, N \cos \varphi}{\partial \lambda} + mN \cos \varphi \cos \gamma \dfrac{\partial \lambda}{\partial \lambda} \\[4mm] \dfrac{\partial^2 X}{\partial \lambda\, \partial \psi} = -\dfrac{\partial^2 Y}{\partial \psi^2} = \sin \gamma \dfrac{\partial m\, N \cos \varphi}{\partial \psi} + mN \cos \varphi \cos \gamma \dfrac{\partial \gamma}{\partial \psi}. \end{array} \right\} \qquad (A.22)$$

After application of (A 22) two linear equations remain, $\partial \gamma / \partial \psi$ and $\partial \gamma / \partial \lambda$ being the unknowns.

The solution gives

$$\frac{\partial \gamma}{\partial \psi} = \frac{1}{mN \cos \varphi} \frac{\partial\, mN \cos \varphi}{\partial \lambda} = \frac{\partial \ln mN \cos \varphi}{\partial \lambda}$$

and

$$\frac{\partial \gamma}{\partial \lambda} = -\frac{1}{mN \cos \varphi} \frac{\partial\, mN \cos}{\partial \psi} = -\frac{\partial \ln mN \cos \varphi}{\partial \psi}.$$

Now

$$\frac{\partial \ln N \cos \varphi}{\partial \psi} = 0$$

and

$$\frac{\ln N \cos \varphi}{\partial \psi} = \frac{1}{M} \frac{\partial N \cos \varphi}{\partial \varphi} = -\sin \varphi.$$

Hence

$$\left.\begin{aligned} \frac{\partial \gamma}{\partial \psi} &= \frac{\partial \ln m}{\partial \lambda} \\[2mm] \frac{\partial \gamma}{\partial \lambda} &= -\frac{\partial \ln m}{\partial \psi} + \sin \varphi. \end{aligned}\right\} \tag{A.23}$$

Referring (A.17) to (A.19) it is found that

$$d\gamma = \frac{\partial \ln m}{\partial \lambda} d\psi - \frac{\partial \ln m}{\partial \psi} d\lambda + \sin \varphi \, d\lambda. \tag{A.24}$$

Substitution of (A.18) and (A.24) into (A.15) gives

$$d\mu = dA_2 = -\frac{\partial \ln m}{\partial \lambda} d\beta + \frac{\partial \ln m}{\partial \psi} d\lambda. \tag{A.25}$$

So far an elementary arc $ds$ of the arc $PQ$ has been considered, having an azimuth $\alpha$ where

$$\tan \alpha = \frac{d\lambda}{d\psi}.$$

The elementary arc $ds_1 = ds$ of a curve at right angles to $PQ$ at $P$ is equal to (azimuth $= \alpha + 270°$)

$$ds_1^2 = N^2 \cos^2 \varphi (d\psi_1^2 + d\lambda_1^2)$$

where

$$\tan(\alpha + 270°) = \frac{d\lambda_1}{d\psi_1} = -\frac{d\beta}{d\lambda}.$$

It may then be noted that (see Fig. A.3)

$$\left.\begin{aligned} d\lambda_1 &= -d\psi \\ d\psi_1 &= d\lambda. \end{aligned}\right\} \tag{A.26}$$

Substitution of (A.26) into (A.25) gives

$$d\mu = dA_2 = \frac{\partial \ln m}{\partial \lambda} d\lambda_1 + \frac{\ln m}{\partial \psi} d\psi_1 = d \ln m. \tag{A.27}$$

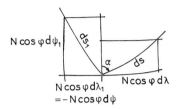

Fig. A.3
Elementary arcs of azimuth $\alpha$ and $\alpha + 270°$

This leads to (A.12)

$$\frac{1}{\rho_{PQ}} = \frac{1}{m} \frac{dm}{dS} = \frac{d \ln m}{dS_{(A_2-90°)}} :$$

the curvature of the projected geodesic is equal to the rate of change of $\ln m$ perpendicular to the geodesic in the direction $(\alpha + 270°) = (\alpha - 90°)$ on the ellipsoid. This is equally true in the projection $(A_2-90°)$.

## A.4  Difference of the chord length $D$ and $S$. Arc to chord correction. Inverse formulae. The deviation of the projected geodesic from the chord

From

$$\frac{1}{\rho_{PQ}} = \frac{d\mu}{dS}$$

follows that the angle between the tangents at $P$ and $Q$ is equal to

$$\mu = \int_0^S \frac{1}{\rho_{PQ}} dS \tag{A.28}$$

This integral is solved by means of an expansion of $1/\rho_{PQ}$ into a MacLaurin series. Putting $\sigma_{PQ} = 1/\rho_{PQ}$ for convenience and indicating the derivatives by primes,

$$\sigma_{PQ} = \sigma_P + \sigma_P' S + \tfrac{1}{2}\sigma_P'' S^2 + \tfrac{1}{6}\sigma_P''' S^3 + \ldots \tag{A.29}$$

where $\sigma_P$ is the curvature at the zero point $P$ of the curve. Hence by integrating each term

$$\mu = \sigma_P S + \tfrac{1}{2}\sigma_P' S^2 + \tfrac{1}{6}\sigma_P'' S^3 + \tfrac{1}{24}\sigma_P''' S^4 + \ldots . \tag{A.30}$$

In order to calculate the difference between the lengths $D$ of the chord $PQ$ and $S$ of the curve $PQ$, as well as the angles $\delta_1$ and $\delta_2$ the coordinate system is rotated about an angle $A_2$ and translated with the origin at $P$. The new abscissa $\overline{X}$ coincides with the tangent to the curve $PQ$ at $P$. Then

$$\left.\begin{aligned}
\cos \mu &= \frac{\mathrm{d}\overline{X}}{\mathrm{d}S} = 1 - \tfrac{1}{2}\mu^2 + \tfrac{1}{24}\mu^4 - \dots \\[2ex]
\sin \mu &= -\frac{\mathrm{d}\overline{Y}}{\mathrm{d}S} = \mu - \tfrac{1}{6}\mu^3 + \dots .
\end{aligned}\right\} \tag{A.31}$$

These expressions become by substitution of (A.30)

$$\left.\begin{aligned}
\frac{\mathrm{d}\overline{X}}{\mathrm{d}S} &= 1 - \tfrac{1}{2}\sigma_P^2 S^2 + \tfrac{1}{2}\sigma_P \sigma_P' S^3 + \dots \\[2ex]
-\frac{\mathrm{d}\overline{Y}}{\mathrm{d}S} &= \sigma_P S + \tfrac{1}{2}\sigma_P' S^2 - \tfrac{1}{6}(\sigma_P^3 - \sigma_P'') S^3 + \dots
\end{aligned}\right\} \tag{A.32}$$

Integration of (A.32) gives

$$\left.\begin{aligned}
\overline{X} &= S - \tfrac{1}{6}\sigma_P^2 S^3 - \tfrac{1}{8}\sigma_P \sigma_P' S^4 - \dots \\
-\overline{Y} &= \tfrac{1}{2}\sigma_P S^2 + \tfrac{1}{6}\sigma_P' S^3 - \tfrac{1}{24}(\sigma_P^3 - \sigma_P'') S^4 \dots
\end{aligned}\right\} \tag{A.33}$$

Now

$$D = \sqrt{\overline{X}^2 + \overline{Y}^2} = S - \tfrac{1}{24}\sigma_P^2 S^3 - \tfrac{1}{24}\sigma_P \sigma_P' S^4 \dots \tag{A.34}$$

(applying the expansion $\sqrt{1+x} = 1 + \tfrac{1}{2}x + \dots$).

The arc to chord corrections in $\delta_1$ and $\delta_2$ are derived as follows

$$\tan (A_2 - A_1) = -\tan \delta_1 = -\frac{\overline{Y}}{\overline{X}}.$$

By (A.33)

$$\tan \delta_1 = \tfrac{1}{2}\sigma_P S + \tfrac{1}{2}\sigma_P' S^2 - \tfrac{1}{24}(\sigma_P^3 - \sigma_P'') S^3 \dots . \tag{A.35}$$

Since

$$\delta_1 = \operatorname{arc} \tan \delta_1 = \tan \delta_1 - \tfrac{1}{3}\tan^3 \delta_1 + \dots$$

by use of (A.35)

$$\delta_1 = \tfrac{1}{2}\sigma_P S + \tfrac{1}{6}\sigma_P' S^2 + \tfrac{1}{24}\sigma_P'' S^3. \tag{A.36}$$

It is easily shown that

$$\delta_2 = \mu - \delta_1 = \tfrac{1}{2}\sigma_P S + \tfrac{1}{3}\sigma_P' S^2 + \tfrac{1}{8}\sigma_P'' S^3 - \dots . \tag{A.37}$$

Some of the terms in the expansions (A.30) for $D$, (A.36) and (A.37) for $\delta_1$ and $\delta_2$ respectively, can be eliminated by the application of a trick similar to that in section A.2 viz. by the introduction of the curvatures $\sigma_{\frac{1}{3}}$ and $\sigma_{\frac{1}{2}}$ at the respective points at $\frac{1}{3}S$ and $\frac{1}{2}S$ of the curve $PQ$.

$$\sigma_{\frac{1}{3}} = \sigma_P + \tfrac{1}{3}\sigma'_P S + \tfrac{1}{18}\sigma''_P S^2 + \ldots \tag{A.38}$$

and

$$\sigma_{\frac{1}{2}} = \sigma_P + \tfrac{1}{2}\sigma'_P S + \tfrac{1}{8}\sigma''_P S^2 + \ldots . \tag{A.39}$$

It is easily corroborated that

$$\mu = \sigma_{\frac{1}{2}} S + \tfrac{1}{24}\sigma''_P S^3 + \ldots \tag{A.40}$$

$$D = S - \tfrac{1}{24}\sigma_{\frac{1}{2}}^2 S^3 + \tfrac{1}{96}\sigma'^2_P S^5 - \ldots \tag{A.41}$$

$$\delta_1 = \tfrac{1}{2}\sigma_{\frac{1}{3}} S + \tfrac{1}{72}\sigma''_P S^3 + \ldots \tag{A.42}$$

and also

$$\delta_2 = \tfrac{1}{2}\sigma_{\frac{2}{3}} S + \ldots . \tag{A.43}$$

The inverse expression of (A.34) is

$$S = D + \tfrac{1}{24}\sigma_P^2 D^3 + \tfrac{1}{24}\sigma_P \sigma'_P D^4 + \ldots \tag{A.44}$$

whence by substituting this into (A.4)

$$\mu = \sigma_P D + \tfrac{1}{2}\sigma'_P D^2 + \ldots . \tag{A.45}$$

Sometimes it is necessary to know the deviation at a point of the projected geodesic from the chord.

Referring to the local coordinate system adopted earlier, with the origin at $P$ and the abscissa coinciding with the tangent to $PQ$ at $P$ (Fig. A.4), then by

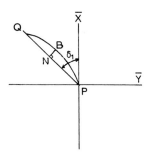

Fig. A.4

simple analytic geometry the deviation $\varDelta_B$ becomes (irrespective of signs)

$$\varDelta_B = \overline{X}_P \sin \delta - \overline{Y}_P \cos \delta. \tag{A.46}$$

Now

$$\sin \delta_1 = \delta_1 - \frac{\delta_1^3}{3!} + \ldots = \tfrac{1}{2}\sigma_P S + \tfrac{1}{6}\sigma_P' S^2 - \tfrac{1}{48}(\sigma_P^3 - 2\sigma_P'')S^3 + \ldots \tag{A.47}$$

using (A.36).
   Similarly

$$\cos \delta_1 = 1 - \frac{\delta_1^2}{2!} + \frac{\delta_1^4}{4!} - \ldots = 1 - \tfrac{1}{8}\sigma_P' S^2 - \tfrac{1}{12}\sigma_P \sigma_P' S^3 \tag{A.48}$$

By substitution of (A.47), (A.48) and (A.38) into (A.46) the deviation in terms of $S$ and $\sigma$ becomes, watching the signs

$$\varDelta_B = \tfrac{1}{2}\sigma_P(SS_B - \sigma_B^2) + \tfrac{1}{6}\sigma_P'(S^2 S_B - S_B^3) + \tfrac{1}{24}\sigma_P''(\sigma^3 S_B - S_B^4) + \ldots \tag{A.49}$$

For any location of $B$ on $PQ$ the deviation is easily obtained e.g. $\varDelta_{\frac{1}{2}}$ by substituting $S_B = \tfrac{1}{2}S$; $\varDelta_{\frac{1}{3}}$ by substituting $S_B = \tfrac{1}{3}S$ into (A.49) etc.; Some of the terms may be eliminated by introducing $\sigma_{\frac{1}{2}}$, $\sigma_{\frac{1}{3}}$, as has been shown eerlier.

## A.5   The meridian convergence

The convergence of the meridian in the projection has been defined as the angle between the abscissa $X$ and the projected meridian. In Figure A.2, $\gamma$ takes the sign of $dY$.
   From (A.21) follows, assuming constancy of $\varphi$

$$\tan \gamma = \frac{\dfrac{\partial X}{\partial \lambda}}{\dfrac{\partial Y}{\partial \lambda}} = \frac{dX}{dY} \tag{A.50}$$

Also, assuming constancy of $\lambda$

$$\tan \gamma = -\frac{\dfrac{\partial Y}{\partial \psi}}{\dfrac{\partial X}{\partial \psi}} = -\frac{dY}{dX} \tag{A.51}$$

On the other hand, keeping $X$ constant (see Fig. A.5)

$$\tan \gamma = -\frac{d\psi}{d\lambda} \tag{A.52}$$

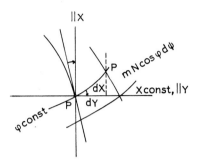

Fig. A.5
The meridian convergence

Referring to formula (A.24)

$$\gamma = \int d\gamma = \int \sin \varphi \, d\lambda + \int \frac{\partial \ln m}{\partial \lambda} d\psi - \int \frac{\partial \ln m}{\partial \psi} d\lambda$$

or

$$\gamma = \lambda \sin \varphi + \ldots \tag{A.53}$$

Thus, at first approximation the convergence at a point $P$ is equal to the product of the difference of the longitude of the reference meridian and that of $P$, and the sine of the latitude of that point, *independent of the type of projection*. The other terms are dependent on the projection, since they include the scale distortion.

For every type of conformal projection the elements $S$, $\sigma$, $D$, $\delta_1$, $\delta_2$, and $\Delta$ can be calculated from the formulae in these last sections, by substituting the appropriate expression for the scale distortion $m$.

## APPENDIX B

## SUMMARY OF MAPPING EQUATIONS

The mapping equations of the projections derived in the previous chapters are given below in terms of rectangular Cartesian coordinates. The coordinates of the origin $X_0$, $Y_0$ correspond to $\varphi_0$, $\lambda_0$ respectively.

1) The gnomonic projection (Section 4.1.1). Spherical formulae.
   Oblique:

$$X = \frac{R\{\cos\varphi_0 \sin\varphi - \sin\varphi_0 \cos\varphi \cos(\lambda - \lambda_0)\}}{\sin\varphi_0 \sin\varphi + \cos\varphi_0 \cos\varphi \cos(\lambda - \lambda_0)}$$

$$Y = \frac{R\cos\varphi \sin(\lambda - \lambda_0)}{\sin\varphi_0 \sin\varphi + \cos\varphi_0 \cos\varphi \cos(\lambda - \lambda_0)}$$

Polar gnomonic projection:      $\varphi_0 \rightarrow 90°$
Transverse gnomonic projection: $\varphi_0 \rightarrow 0°$

2) The stereographic projection (Section 4.1.2). Spherical formulae.
   Oblique:

$$X = \frac{2R\{\cos\varphi_0 \sin\varphi - \sin\varphi_0 \cos\varphi \cos(\lambda - \lambda_0)\}}{1 + \sin\varphi_0 \sin\varphi + \cos\varphi_0 \cos\varphi \cos(\lambda - \lambda_0)}$$

$$Y = \frac{2R\cos\varphi \sin(\lambda - \lambda_0)}{1 + \sin\varphi_0 \sin\varphi + \cos\varphi_0 \cos\varphi \cos(\lambda - \lambda_0)}$$

Polar: $\varphi_0 \rightarrow 90°$
Transverse or equatorial: $\varphi_0 \rightarrow 0°$

3) The orthographic projection (Section 4.1.3). Spherical formulae.
   Oblique:

$$X = R\{\cos\varphi_0 \sin\varphi - \sin\varphi_0 \cos\varphi \cos(\lambda - \lambda_0)\}$$

$$Y = R\cos\varphi \sin(\lambda - \lambda_0)$$

Polar: $\varphi_0 \rightarrow 90°$
Transverse: $\varphi_0 \rightarrow 0°$

4) The azimuthal equidistant projection (Postel) (Section 4.1.4). Spherical formulae.
Local coordinate system.

$$X = \text{arc}\,\delta\,\cos\alpha$$

$$Y = \text{arc}\,\delta\,\sin\alpha$$

For expressions of $X$ and $Y$ as functions of $(\varphi, \lambda)$, apply a transformation with (4.4) (4.5) and (4.6) (p. 59).

5) The azimuthal equivalent projection (Lambert (Section 4.1.5). Spherical formulae.
Local coordinate system.

$$X = 2\,R\,\sin\tfrac{1}{2}\,\delta\,\cos\alpha$$

$$Y = 2\,R\,\sin\tfrac{1}{2}\,\delta\,\sin\alpha$$

See also under 12) below.

6) The Cassini–Soldner Projection (normal) (Section 4.2) $X = X_D + \tfrac{1}{2}N$ $(\Delta\lambda\,\cos\,\varphi)^2\,\tan\,\varphi + \tfrac{1}{24}N\,(\Delta\varphi\,\cos\,\varphi)^4\,\tan\,\varphi\,(5 - \tan^2\varphi)$. $Y = N\,(\Delta\varphi\,\cos\,\varphi)$ $- \tfrac{1}{6}N\,(\Delta\varphi\,\cos\,\varphi)^3\,\tan^2\varphi - \tfrac{1}{120}N\,(\Delta\varphi\,\cos\,\varphi)^5\,\tan^2\varphi\,(8 - \tan^2\varphi)$

7) The normal conformal conical Lambert projection (Section 5.3.2)
Ellipsoidal formulae;
a) One standard parallel: $\varphi_0$

$$X = N_0\,\cot\varphi_0 - \rho\,\cos(\lambda\,\sin\,\varphi_0)$$

$$Y = \rho\,\sin(\lambda\,\sin\,\varphi_0)$$

$$m = \rho\,\frac{\sin\varphi_0}{N\,\cos\varphi}$$

$$\rho = N_0\,\cot\varphi_0 \left\{ \frac{\tan(45° - \tfrac{1}{2}\varphi)\left(\dfrac{1 + \varepsilon\sin\varphi}{1 - \varepsilon\sin\varphi}\right)^{\tfrac{1}{2}\varepsilon}}{\tan(45° - \tfrac{1}{2}\varphi_0)\left(\dfrac{1 + \varepsilon\sin\varphi_0}{1 - \varepsilon\sin\varphi_0}\right)^{\tfrac{1}{2}\varepsilon}} \right\}^{\sin\varphi_0}$$

b) Two standard parallels: $\varphi_1$ and $\varphi_2$

$$X = \rho_0 - \rho\,\cos(\lambda\,\sin\,\varphi_0)$$

$$Y = \rho\,\sin(\lambda\,\sin\,\varphi_0)$$

$$m = \rho\,\frac{\sin\varphi_0}{N\,\cos\varphi}$$

$$\rho = C \left\{ \tan(45° - \tfrac{1}{2}\varphi) \left( \frac{1 + \varepsilon \sin \varphi}{1 - \varepsilon \sin \varphi} \right)^{\frac{1}{2}\varepsilon} \right\}^{\sin \varphi_0}$$

$\rho_0$ is the polar ray through the centre $X_0$, $Y_0$ of the map.

$$C = \frac{N_1 \cos \varphi_1}{\sin \varphi_0 \left\{ \tan(45° - \tfrac{1}{2}\varphi_1) \left( \frac{1 + \varepsilon \sin \varphi_1}{1 - \varepsilon \sin \varphi_1} \right)^{\frac{1}{2}\varepsilon} \right\}^{\sin \varphi_0}} =$$

$$= \frac{N_2 \cos \varphi_2}{\sin \varphi_0 \left\{ \tan(45° - \tfrac{1}{2}\varphi_2) \left( \frac{1 + \varepsilon \sin \varphi_2}{1 - \varepsilon \sin \varphi_2} \right)^{\frac{1}{2}\varepsilon} \right\}^{\sin \varphi_0}}$$

$$\sin \varphi_0 = \frac{\ln N_1 \cos \varphi_1 - \ln N_2 \cos \varphi_2}{\ln \tan(45° - \tfrac{1}{2}\varphi_1) \left( \frac{1 + \varepsilon \sin \varphi_1}{1 - \varepsilon \sin \varphi_1} \right)^{\frac{1}{2}\varepsilon} - \ln \tan(45° - \tfrac{1}{2}\varphi_2) \left( \frac{1 + \varepsilon \sin \varphi_2}{1 - \varepsilon \sin \varphi_2} \right)^{\frac{1}{2}\varepsilon}}$$

8) The Mercator conformal projection (Section 5.3.3). Ellipsoidal formulae.

$$X = a \ln \tan(45° + \tfrac{1}{2}\varphi) \left( \frac{1 - \varepsilon \sin \varphi}{1 + \varepsilon \sin \varphi} \right)^{\frac{1}{2}\varepsilon}$$

$$Y = a\lambda$$

$$m = \frac{a}{N \cos \varphi} = 1 + \frac{(1 - \varepsilon^2) X^2}{2 a^2}$$

($a$ is the radius of the equatorial circle)

9) The polar stereographic projection (Section 5.3.3). Ellipsoidal formulae.

$$X = \rho \cos \lambda$$

$$Y = \rho \sin \lambda$$

$$\rho = 2a(1 + \varepsilon)^{-\frac{1}{2}(1 - \varepsilon)} (1 - \varepsilon)^{-\frac{1}{2}(1 + \varepsilon)} \tan(45° - \tfrac{1}{2}\varphi) \left( \frac{1 + \varepsilon \sin \varphi}{1 - \varepsilon \sin \varphi} \right)^{\frac{1}{2}\varepsilon}$$

$$m = 1 + \frac{(1 + \varepsilon)^{(1 - 2\varepsilon)} (1 - \varepsilon)^{(1 + 2\varepsilon)}}{4 a^2} (X^2 + Y^2)$$

For the corresponding spherical formulae of 6 a) and b); 7, and 8 see Section 5.3.4.

10) The oblique Mercator projection, (Section 5.3.5) Spherical formulae.

$$X = \tfrac{1}{2} R \ln \frac{1 + \sin \varphi \, \sin \varphi_0 + \cos \varphi \, \cos \varphi_0 \, \cos \lambda}{1 - \sin \varphi \, \sin \varphi_0 - \cos \varphi \, \cos \varphi_0 \, \cos \lambda}$$

$$Y = R \arctan \frac{\sin \lambda \, \cos \varphi}{\cos \varphi \, \sin \varphi_0 \, \cos \lambda - \cos \varphi_0 \, \sin \varphi}$$

The Transverse Mercator projection: $\varphi_0 \to 0$.

11) The normal equivalent Albers projection (Section 6.3.2) Spherical formulae.
   a) One standard parallel: $\varphi_0$

$$X = \frac{R}{\sin \varphi_0} \cos(\lambda \sin \varphi_0) \, \{\cos\varphi_0 - \sqrt{1 + \sin^2 \varphi_0 - 2 \sin \varphi \, \sin \varphi_0}\}$$

$$Y = \frac{R}{\sin \varphi_0} \sin(\lambda \sin \varphi_0) \, \sqrt{1 + \sin^2 \varphi_0 - 2 \sin \varphi \, \sin \varphi_0}$$

   b) Two standard parallels: $\varphi_1$ and $\varphi_2$

$$X = \rho_0 - \rho \cos\{\tfrac{1}{2}(\sin \varphi_1 + \sin \varphi_2)\lambda\}$$

$$Y = \rho \sin\{\tfrac{1}{2}(\sin \varphi_1 + \sin \varphi_2)\}$$

$$\rho = \sqrt{\rho_1^2 + \frac{4 R^2 (\sin \varphi_1 - \sin \varphi)}{\sin \varphi_1 + \sin \varphi_2}} =$$

$$= \sqrt{\rho_2^2 + \frac{4 R^2 (\sin \varphi_2 - \sin \varphi)}{\sin \varphi_1 + \sin \varphi_2}}$$

$$\rho_1 = \frac{2 R \cos \varphi_1}{\sin \varphi_1 + \sin \varphi_2} \qquad \rho_2 = \frac{2 R \cos \varphi_2}{\sin \varphi_1 + \sin \varphi_2}$$

$\rho_0$ is the polar ray through the centre $X_0$, $Y_0$ of the map.

12) Lambert's cylindrical normal equivalent projection (Section 6.33). Spherical formulae.

$$X = R \sin \varphi$$

$$Y = R\lambda \qquad \text{(radians)}$$

13) Lambert's polar azimuthal equivalent projection. Spherical formulae.

$$X = 2R \sin(45° - \tfrac{1}{2}\varphi)\cos\lambda$$

$$Y = 2R \sin(45° - \tfrac{1}{2}\varphi)\sin\lambda$$

14) Bonne's normal equivalent pseudo-conical projection (Section 6.4.2) Spherical formulae.

$$X = R \cot\varphi_0 - \rho \cos\left(\frac{R\cos\varphi}{\rho}\lambda\right)$$

$$Y = \rho \sin\left(\frac{R\cos\varphi}{\rho}\lambda\right)$$

$$\rho = R \cot\varphi_0 - R(\varphi - \varphi_0)$$

15) Sanson-Flamsteed projection. Spherical formulae.

$$X = R\varphi$$

$$Y = \lambda R \cos\varphi$$

16) Werner projection. Spherical formulae.

$$X = R(90-\varphi)\cos\frac{\cos\varphi}{90-\varphi}\lambda$$

$$Y = R(90-\varphi)\sin\frac{\cos\varphi}{90-\varphi}\lambda$$

APPENDIX C

CONSTANTS

Base of Naperian logarithms $\qquad e = 2.718\ 281\ 83$

$\qquad \log e = 0.434\ 294\ 48$

Modulus of common logarithms $\qquad M = 0.434\ 294\ 48$

$\qquad \log M = 9.637\ 784\ 31 - 10$

$\pi = 3.141\ 592\ 65$

$\log \pi = 0.497\ 149\ 87$

LENGTHS

The international yard equals 0.9144 metre (after July 1, 1959)

1 metre = 39.370 078 8 inches = 3.280 839 9 feet = 1. 09 361 33 yards

1 inch = 25.4 millimetres = 0.025 4 metres

1 foot = 0.304 8 metres

1 statute mile = 1 609.344 0 metres = 1.609 344 0 kilometres = 5280 feet

1 yard = 3 feet = 36 inches

1 Nautical Mile a) International 1852 metres

b) British $\qquad$ 6080 ft = 1853.184 metres

AREAS

1 sq. kilometre = $10^6$ sq. metres

1 hectare = $10^4$ sq. metres

1 sq. inch = $6.451\ 6 \times 10^{-4}$ sq. metres

1 sq. foot = $9.290.304 \times 10^{-2}$ sq. metres

1 sq. yard = 0.836 127 36 sq. metres

1 acre = 4840 sq. yards

1 sq. mile = 640 acres

DEGREES, RADIANS

$$\rho = 57.295\ 78° \qquad = 3\ 437.746\ 8' \qquad = 206\ 264.8'' = \left(\frac{1}{\sin 1''}\right)''$$

# BIBLIOGRAPHY

The bibliography is listed in alphabetical order of authors. The title of a book or an article is given first, then follows the name of the publisher, place and date; or the title of the periodical, date and possibly page numbers respectively.

The language – if not English – is indicated by a capital following the title: D = Dutch; F = French; G = German.

The list below is by no means exhaustive.

1. ADAMS, O. S., General theory of equivalent projections, U.S. Coast and Geodetic Survey, Washington D.C. Special Publication no. 236.
2. ADAMS, O. S., General theory on the Lambert conformal conic projection, U.S. Coast and Geodetic Survey, Washington D.C. Special Publications no. 52 and 53.
3. ADLER, R. K., REILLY, J. P. and SCHWARTZ, C. R., A generalized system for the evaluation and automatic plotting of map projections, *The Canadian Surveyor* **XXII**, no. 5, 1968.
4. BALCHIN, W. G. V., Air maps, *Empire Survey Review*, Oct. 1947.
5. CRAIG, J., Theory of map projections, *Survey of Egypt*, Cairo, 1910.
6. DEETZ, Ch. H. and ADAMS, O. S., Elements of map projection, U.S. Coast and Geodetic Survey, Washington D.C. Special Publication no. 68.
7. DRIENCOURT, L. and LABORDE, J., *Traité des projections des cartes géographic à l'usage des cartographes et des géodésiens*, (4 fascicules) (F) Hermann et Cie, Paris, 1932.
8. GOUSSINSKY, B., *On the classification of map projections*, Empire Survey Review **11**, 1951.
9. GROSSMANN, W., *Geodätische Rechnungen und Abbildungen in der Landesvermessung* (G), Wittwer, Stuttgart, 1964.
10. HOSCHEK, J., *Mathematische Grundlagen der Kartographie* (G), Bibliographisches Institut no. 443/443a, Mannheim/Zürich, 1968.
11. HOTINE, M., The orthomorphic projection of the Spheriod, Empire Survey Review, Oct. 1946; Jan.–April–July–Oct., 1947.
12. LAUBSCHER, A. L., *A basic investigation of perspective map projections*, M.Sc. thesis, Ohio State University, 1965.
13. MALING, D. H., Suitable projections for maps of the visible surface of the moon, *The Cartographic Journal*, Dec. 1965.
14. MALING, D. H., "The terminology of map projections", *International Yearbook of Cartography* **VIII**, 1968.
15· McBRYDE, F. W. and THOMAS, P. D., Equal area projections for world Statistical maps, U.S. Coast and Geodetic Survey, Washington D.C. Special Publication no. 245.
16. MERKEL, H., *Grundzüge der Kartenprojektionlehre* (G), Deutsche Geodätische Kommission, Series A no. 17, Munich 1958.
17. MYERS, D., *An investigation of map projections for use in lunar cartography*, M.Sc. thesis, Ohio State University, 1969.
18. ROBINSON, A. H., The use of deformational data in evaluating world map projections, *Annals of the Assoc. of American Geographers* **XLI** no. 1, 1951.
19. ROGGEVEEN, C. and MASSINK, H. R., Een richting- en afstandskaart (D), Tijdschrift voor Kadaster en Landmeetkunde, 1949, 225.
20. Royal Netherlands Geographic Society, *Map projections* (D), Workshop, 1968.
21. SCHOLS, Ch. M., *La courbure de la projection de la ligne geodésique* (F), Annales de l'École Polytechnique de Delft, 1886.
22. SCHOLS, Ch. M., *Studiën over Kaartprojectiën* (F), Drabbe, Leiden, 1882.

23. STEERS, J. A., *An introduction to the study of map projections*, University of London Press, London, 1956.
24. STRANG VAN HEES, G. L., *Kaart-projecties* (D), The Geodetic Institute, Delft, 1963.
25. STRASSER, G., *Ellipsoidische Parameter der Erdfigur* (G), Deutsche Geodätische Kommission Series A no. 19, Munich, 1957.
26. THOMAS, P. D., Conformal projections in Geodesy and Cartography, U.S. Coast and Geodetic Survey, Washington D.C. Special Publication no. 251.
27. TISSOT, M. A., *Mémoire sur la représentation des surfaces et les projections des cartes géographiques* (F), Gauthier-Villars, Paris, 1881.
28. VERSTELLE, J. Th., *Kaartprojecties beschouwd uit een hydrografisch oogpunt* (D), Hydrographic Service of the Royal Dutch Navy, National Press, The Hague, 1951.
29. WEATHERBURN, C. E., *Differential geometry of three dimensions*, London, 1952.
30. ZÖPRITZ, K. and BLUDAU, A., *Leitfaden der Kartenentwurfslehre* (G), Teubner, Leipzig, 1912.

# INDEX